D0778680

Books of Merit

ALSO BY JOHN W. MOFFAT

Reinventing Gravity

Einstein Wrote Back

John W. Moffat

EINSTEIN

WROTE BACK

My Life

in Physics

THOMAS ALLEN PUBLISHERS

TORONTO

Library and Archives Canada Cataloguing in Publication

Moffat, John W., 1932–
    Einstein wrote back / John W. Moffat.

ISBN 978-0-88762-615-9

1. Moffat, John W., 1932– .   2. Moffat, John W., 1932– —Friends and associates.
3. Physicists—Canada—Biography.   4. Physics—History—20th century.   I. Title.

QC16.M64A3 2010   530.092   C2010-903799-5

Editor: Janice Zawerbny
Jacket design: Michel Vrána
Jacket photo of Einstein letter: Copyright Albert Einstein Archives,
    Hebrew University of Jerusalem, Israel.

Published by Thomas Allen Publishers,
a division of Thomas Allen & Son Limited,
145 Front Street East, Suite 209,
Toronto, Ontario M5A 1E3 Canada

www.thomasallen.ca

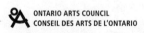

ONTARIO ARTS COUNCIL
CONSEIL DES ARTS DE L'ONTARIO

Canada Council
for the Arts

The publisher gratefully acknowledges the support of The Ontario Arts Council for its publishing program.

We acknowledge the support of the Canada Council for the Arts, which last year invested $20.1 million in writing and publishing throughout Canada.

We acknowledge the Government of Ontario through the Ontario Media Development Corporation's Ontario Book Initiative.

We acknowledge the financial support of the Government of Canada through the Canada Book Fund for our publishing activities.

1 2 3 4 5   14 13 12 11 10

Printed and bound in Canada

Again to Patricia,
whose dedication made this book possible,

and to Sandra, Tina,
Derek and Tessa

CONTENTS

Prologue   1

CHAPTER 1
Child of War   7

CHAPTER 2
Two Paths Diverged   19

CHAPTER 3
Niels Bohr   33

CHAPTER 4
Albert Einstein   43

CHAPTER 5
Erwin Schrödinger   59

CHAPTER 6
Fred Hoyle   73

CHAPTER 7
The Einstein Fest   93

CHAPTER 8
Wolfgang Pauli   105

CHAPTER 9
Paul Dirac   119

CHAPTER 10
Abdus Salam   139

CHAPTER 11
Imperial College London   151

CHAPTER 12
Baltimore   161

CHAPTER 13
CERN and Particle Physics   179

CHAPTER 14
Princeton and Oppenheimer   191

CHAPTER 15
Toronto   201

Epilogue   227

Acknowledgements   236

Index   237

PROLOGUE

T HROUGH the large picture window in my office at the
Perimeter Institute for Theoretical Physics in Waterloo,
Ontario, I have a view of Silver Lake in a nearby park. While
pondering the mysteries of the universe, I often watch the swans
gliding back and forth across the lake, and the children in the play-
ground on the other side.

The Institute—affectionately called "PI"—was founded by Mike
Lazaridis, the inventor of the BlackBerry, whose company, Research
in Motion (RIM), is headquartered in Waterloo. Lazaridis has contributed generous funds to creating PI, where over one hundred
theoretical physicists from around the world spend their time following impractical dreams: searching for quantum gravity, understanding the beginnings of the universe and probing the quantum
nature of matter. As the name of the Institute implies, we physicists
who work there are out on the "perimeter" or the cutting edge of
fundamental physics. PI is an ideal place for me to be, as since
my unusual beginning in physics, I have been mostly involved in
searching for new ways to come to a fundamental understanding
of the universe. This kind of physics is often referred to as "outside the box." Those of us who practise it think about physics in an

unconventional way, attempting to view old problems in novel ways, or to ask unusual questions that may bear fruit in unexpected ways. Yet ultimately we always hope to relate our speculative theories to the reality of nature by comparing the predictions of our theories with experiments and observations.

Occasionally I stare out my window at Silver Lake and think about the bizarre way my life unfolded, eventually leading me to this place. In my peripatetic and traumatic childhood in Denmark, England and Scotland during and after the Second World War, I showed little aptitude for mathematics and science—so little, in fact, that I was not even allowed to enter university. Instead, I set my sights on becoming an abstract painter, an almost impossible career choice in the immediate postwar years. But then something peculiar happened to me to change drastically the course of my life. Within little more than a year, I vaulted from working at odd jobs in Copenhagen—window cleaner, delivery boy, mail sorter—to entering the Ph.D. program in physics at Trinity College, Cambridge.

How did this happen? What did it mean? Colleagues as well as my family have often encouraged me to write about my early life and the unusual way that I entered physics—and to put down on paper the many anecdotes with which I had regaled them, about the famous physicists of the twentieth century that I had the good fortune to meet. When I ask myself how I became a physicist in the first place, and how I managed to remain outside the box of conventional physics, working on truly fundamental questions of nature throughout so much of my career, my thoughts keep returning to the difficulties of my childhood, the love of beauty that inspired my first career as an artist and the influence of those giants of physics, whose kindness and help encouraged me on my way.

Physicists explore the nature of the universe, from its farthest edges to the smallest constituents of matter. In the twentieth century, with amazing improvements in telescopes and many space missions, we were able to expand our understanding of the evolution of the universe back almost to its beginning. This is a remarkable development in the history of science, because from the Greeks up until the beginning of the twentieth century, our astrophysical investigations were restricted to the much smaller universe of our solar system and our galaxy. We have also made great strides in penetrating the universe of the very small, and gradually the mysteries of the structure of matter are being revealed to us. Cracking the quantum code of matter is only possible through extraordinarily high-energy accelerators such as the Large Hadron Collider (LHC) at the largest particle physics laboratory in the world, at CERN (European Organization for Nuclear Research), near Geneva, which began operation in 2010. We are on the threshold of exciting new discoveries in the realm of particle physics, which will help unravel the mysteries of the nature of matter.

Theoretical physicists attempt to build models of nature based on mathematics, and experimental physicists provide the data that can test the ideas and models proposed by the theoretical physicists. In practice there is an interplay between theory and experiment. Often, successful research in theoretical physics starts with a well-grounded knowledge of experimental data, building up from this data into a theory. Another important area of physics is industrial physics, where developing new technologies eventually leads to advances in computers, televisions, cellphones, medical diagnostics and many other electronic applications. All of these devices grew out of abstract theoretical ideas and their subsequent verification by experimental physics.

From the beginning of my studies, I wanted to become a theoretical physicist. I was fascinated by the intellectual adventure of trying to figure out how the universe worked, using its language of mathematics. I was attracted to the double-edged approach that theoretical physicists must take, combining a dreamer's awe of nature's

3

inner workings with the rigour of having to verify one's ideas and models of nature with data, whether from telescopes or particle colliders. I also felt more comfortable working mainly on my own, as most theoretical physicists do, than working in the large teams that constitute so much of contemporary experimental physics.

To me, the only physics worth doing is outside-the-box, non-mainstream physics, for that is how our understanding of nature moves forward. Of course, I have, like most other physicists, spent time in my career working out the details of someone else's theory, calculating the consequences of someone else's ideas. But that, to me, is not exciting and groundbreaking work. Niels Bohr, Albert Einstein, Erwin Schrödinger, Paul Dirac and others described in this memoir all worked on non-mainstream physics as a matter of course. They all broke through the boundaries of what consti-tuted the conventional paradigm in physics in their day. Perhaps my interactions with them as a young student steered me in this direction of always aiming to do the kind of physics that challenges the conventional wisdom.

Albert Einstein, in particular—and the letters we exchanged over several months—opened the doors for me into the academic world. Einstein was always an iconoclastic physicist, and his revo-lutionary ideas were not immediately accepted by the physics com-munity. Indeed, in some cases, such as his interpretation of light as photon particles, it took several years before his ideas were incor-porated into mainstream physics, and became part of the early revolutionary development of quantum mechanics.

From the very beginning of my research career as a student at Trinity College, Cambridge, in the early 1950s, I attempted, like Einstein, Paul Dirac, Werner Heisenberg and other well-known twentieth-century physicists, to get at the heart of the fundamental issues in physics. In my first three papers, published while I was a student, I devoted myself to modifying Einstein's gravity theory. In this effort I was actually following in Einstein's footsteps, for after developing his great theory of gravity, general relativity, in 1915, he sought a unified field theory of gravity and electromagnetism, which necessitated modifying general relativity.

I always say that to achieve success in fundamental, theoretical physics, one must be childishly optimistic, possess a thick skin and live a long life. This memoir is an attempt to trace the origins of the desire to work on non-mainstream, fundamental science in my own life. I hope that this narrative will entertain you, that you will enjoy this journey into the company of the giants of modern physics who were my mentors.

1

————

CHILD OF WAR

MY PARENTS MARRIED three weeks after they met, speaking barely a word of each other's language.

My father, George Moffat, was born in 1907 in Glasgow, Scotland. As he grew up there, he played musical instruments from a young age, beginning with a classical-piano teacher who had been a pupil of Franz Liszt. My father also took up the trumpet in the Scottish Boys' Brigade, won the all-Scotland championship in coronet playing and taught himself to play the accordion. In addition to his musical talents, he was a successful artist, and at age seventeen won a scholarship to go to Rome to study painting. His father forbade him to go, however, for he wanted my father to work in his tailoring business in Glasgow. But my father left home and joined a band in England, playing the trumpet. The band toured around Europe just before the Second World War and ended up at the La Scala nightclub in Copenhagen, where my mother, Esther Winther, a local girl, was working as a chorus dancer. They immediately fell into a heady romance.

My mother learned that she was pregnant while my father was touring with his band in Norway. Her gynecologist was very surprised that she had become pregnant, for she suffered from a serious condition that he had told her would prevent her from ever having a child. Throughout her life, my mother always spoke of her pregnancy and my birth as a miracle. She never became pregnant again.

In 1938, when I was six, my father foresaw that the Nazis would invade Denmark. Since he was still an alien, a British citizen holding a British passport, the Nazis would have detained all three of us and put us in a camp. So we moved to Britain and, my parents believed, to greater safety.

When war broke out in September 1939, the entertainment industry in England closed overnight because of the widespread fear that the Germans would bomb London. My father took a job as a truck driver for a pharmaceutical company, which was dangerous work, driving around London at night in the blackouts. He then worked in the intelligence service, starting by censoring servicemen's letters from abroad.

Concerned for my safety, given the ominous signs of the coming bombings, in late 1939, when I was seven years old, my parents evacuated me from London, putting me alone on a double-decker bus to Glasgow, where my grandparents lived. During the bus trip I was sustained by the sandwiches my mother had packed, and the kind reassurances of the bus driver. I lived with my grandparents and my aunt Rhoda for a year as an evacuee, attending a nearby school. Not surprisingly, I did not do particularly well at school that year in Glasgow.

In fact, my grandparents and my aunt Rhoda could see that I was not thriving in the absence of my mother and father, so despite the danger of the blitzes as the German raids on southern England intensified, they sent me back to London alone by train. We soon left for Bristol, where my father worked as an intelligence officer searching incoming ships for Nazi spies.

By 1940, the bombing had increased, and in August of 1940 the Battle of Britain began, lasting through September. This was a prelude to the Nazis' planned invasion of Britain. In an operation that we later learned was code-named "Sea Lion," they intended to land on the Kent and Sussex beaches. The heightened bombing in the summer was designed to control the English Channel so that the British navy would not be able to destroy the German barges that would bring tanks and troops during the invasion.

In Bristol we lived in two rented rooms in a house that was not far from the British Aeroplane Company in Filton that made the famous Spitfire and Hurricane fighter planes, which enabled the British eventually to win the Battle of Britain. The German pilots who flew in at night were well aware of the location of this facility. After London, Liverpool and Birmingham, Bristol was the fourth most heavily bombed city in Britain during the war. I would lie awake until eleven at night, waiting for the sirens to start wailing, heralding another bombing attack. I would issue a silent prayer to a God unknown that I would survive the night. My parents and I often had to huddle under the stairwell, trying to sleep on makeshift mattresses in that supposedly safest place in the house, which could protect us from a direct hit.

Starting at seven in the morning, when children were on their way to school, the air raid siren would wail again, and waves of German bombers would roar over Bristol. As a young child, I walked alone through streets destroyed by incendiary and high-explosive bombs, my shoes crunching on broken glass. One morning as I walked through the streets filled with rubble, I picked up a pamphlet, dropped as propaganda by German bombers the night before to persuade the English to capitulate. It showed a picture of a child with the top half of her head blown off, a victim of the German bombs. I stuffed it into my school satchel to show my parents later.

Holding tightly to my satchel and gas mask box, I managed every day to reach my class, which was deep underground in a cavernous air raid shelter. During the afternoons, the class would be brought up outside for some fresh air. We often sat on sandbags, eating the lunches our mothers had packed, and watched the dog-fights between the Spitfires and Messerschmitts up in the blue sky. The silver-and-grey Spitfires and black Messerschmitts traced out white contrails as they circled one another, and we heard the *rat-a-tat-tat* of machine guns. On the way home from school, invariably the sirens would wail again. I often would have to knock on the doors of strangers' houses and stay with them until the all-clear siren sounded.

In the late summer of 1940, at the height of the Battle of Britain, my parents and I took a holiday in Weston-super-Mare, a small town on the west coast of England, in an attempt to have a respite from the day-and-night bombings in Bristol. One afternoon we were walking along the boardwalk, eating shrimp from paper cups and viewing the bathers sunning on the beach. Suddenly there was a roaring noise above us. Looking up, I saw two black Messerschmitt fighter bombers passing directly over us, and I got a glimpse of one of the German pilots in his black helmet and goggles. They were being pursued by two Spitfires, and in order to lighten their load to make a hasty escape, the Messerschmitts dropped six whistling bombs on the beach.

I heard the shriek of the whistling bombs as they fell, and then the hollow booms as they detonated deep inside the mud of the beach. Although the mud dampened the effect of the blast, every-one who had been bathing on the beach vanished. The blast blew my parents and me across the road adjacent to the boardwalk. I landed in a garden on my back, opened my eyes and stared at the blue sky, and there was a loud ringing in my ears. The blood was pouring out of my nose, and I felt a terrible tightness and pain in my chest. But otherwise I did not appear to be seriously hurt. That was the amazing phenomenon of the blast, which could lift you as if by a giant's hand and deposit you in a garden without serious physical damage. In a daze, I got up, and soon discovered my parents in the same garden, on all fours, attempting to stand up, also suffering from nosebleeds and chest pains. They also were not seriously hurt by the blast.

At the time, I was somehow able to suppress the horror of our experiences during the war, and carry on day by day. However, about a year after the bombings in Bristol and Weston-super-Mare, I began suffering from what is now called post-traumatic stress disorder. I began getting severe nightmares and panic attacks. Even today I still occasionally experience panic attacks, generally when I am visiting Europe.

*

From Bristol my father was posted to the isolated farming and port town of Stranraer in western Scotland, where he was in charge of port security. Travelling by rail at this time during the war was an arduous and dangerous experience. All our belongings were in two suitcases during this move, and my father carried his precious trumpet in a black case. The trip to the western coast of Scotland took more than three days. We had to change trains because the Germans had bombed the railways connecting England and Scotland. I tried to sleep at night in the train carriage, which was stuffy with the smell of cigarette smoke and the anti-lice chemical odour of the British soldiers' uniforms. The soldiers were with us in the train because military personnel took precedence over civilians in wartime transportation.

On the second night of the journey, near the industrial town of Doncaster in northern England, German bombs fell all around us as we approached the main railway station. The train stopped and we all had to get out and walk. During the ensuing panic, we lost our luggage, including my father's trumpet. We walked, terrified, along the tracks as the bombs fell over the city, and finally got to the railway station, where after some hours we were able to board another train and continue our journey to Scotland.

At dawn of the third day, a cold, foggy morning, we arrived at the railway station at Stranraer. My parents were upset by the experiences of the trip and had a loud argument about whose fault it was for losing all our belongings on the way. My father managed to get hold of a military jeep, and at six o'clock in the morning he drove us to a house that he randomly chose on a dismal street in Stranraer, and he hammered on the door. A sleepy-looking woman with curlers in her hair appeared in her dressing gown. My father announced that he had a war permit that meant he could requisition a room for us for a few nights. The poor woman looked astonished, but then rallied, as people did during the war, and welcomed us in. She escorted me into a bedroom where her three adult daughters were getting up and were in the process of dressing. A small bed in a corner would be mine for the next three nights, as I shared the room with the young ladies.

Despite our initial difficulties in Stranraer, we soon felt grateful that we had left Bristol. My father heard from a colleague who had lived near us in Bristol that the house in which we had been renting rooms had been destroyed about a week after we left for Scotland and our eighty-year-old landlady had been killed. She was in the habit of opening her bedroom window during the bombing barrages and shaking her fist at the German bombers droning overhead. My father's colleague told him that a bomb falling near her house had blown her head off.

After a few days we moved from our emergency lodgings to an old house outside Stranraer owned by two elderly sisters. Some weeks went by before it could be arranged for me to go to school, and during that time of freedom I became friends with a boy who lived on a nearby farm, and he and I roamed through the woods during the day hunting rabbits with slingshots. Once I started school, I often spent time in the late afternoons by myself on the deserted beaches, which were covered with barbed wire and other anti-invasion devices. One night not long after I had started school, a British bomber torpedoed a German troop ship that was sailing between the west coast of Scotland and Ireland. For days after the attack, I watched the bodies of German soldiers washing up on shore.

In Stranraer harbour, there was a Royal Air Force base, from which planes went out on patrol at night on the North Sea and the Atlantic. I spent time at the base and became a sort of mascot of the air force pilots. Some of the young pilots and crew that I got to know didn't return from their patrols. I also spent time after school with the soldiers who occupied an anti-aircraft position not far from the house we were living in, and they would give me chocolate.

It was difficult sharing accommodations with strangers during the war. My mother, an excellent cook who was proud of her kitchen, often got into arguments with our landladies. And so, it was not long before we left the house outside Stranraer and moved into a house in town, renting rooms on the top floor with our own kitchen and bathroom. The house was owned by a dour, unpleasant older man. My father was usually away every night at the port, boarding vessels that came in from southern Ireland. He was searching for Nazi spies who would have entered Ireland from U-boats and then stowed away in fishing boats and smaller vessels coming in from Dublin and Belfast.

One morning, when my father returned from his shift at the port, and I was in the kitchen with my mother, who was preparing his breakfast, he announced that he had joined the army as a private.

He felt that he was not contributing enough to the war effort at the port. My mother was shocked by this decision and couldn't understand why he would do this at the age of thirty-four. He soon started a rigorous six-month program of commando training in an intelligence unit of the army in the south of England. My mother and I stayed on for two months in Stranraer and then moved to Glasgow, where we lived briefly with my grandparents and my aunt Rhoda until we found a flat of our own to rent. In Glasgow, my mother had a government job censoring the letters written home by British soldiers on duty abroad. She had become fluent in reading and writing English during the years she had been married to my father.

My father came back to Glasgow on leave once during his training. When my mother and I met him at the railway station, I barely recognized him getting off the train, carrying his duffle bag and rifle on his shoulders. He had lost so much weight during his strenuous training that he was a shadow of the father I had known. During his leave, and while my parents were away from the small flat, I enjoyed playing with his service rifle and revolver. At one point I overheard my father telling my mother that because he spoke Danish, he had been ordered to be part of a spy operation to be dropped into occupied Denmark by parachute. Fortunately for my father, the operation was cancelled, for his chances of surviving this adventure were low.

In Glasgow I was put into a first-rate educational institution, Allan Glen's School, where unfortunately I did poorly academically, except for the chemistry class, where for some inexplicable reason I always got the top mark. My mathematical abilities did not impress my mathematics teacher, and my physics teacher considered me an abysmal failure. However, I discovered that I had inherited my father's talent for drawing, and won an all-Glasgow drawing competition, which greatly pleased my art teacher.

After his training, my father was promoted to corporal and was stationed in Glasgow. Since the flat my mother had rented was too small for all three of us, my father rented rooms in a house in the western part of Glasgow. Unfortunately, not long after moving in, my mother had yet another altercation with the unpleasant landlady with whom she shared the kitchen. My mother objected to the mouse tracks in the fat in the frying pans, which the landlady had not washed. And so, one night, my father came in and pulled me out of bed and said that we were leaving. We took off in the dark in a military jeep with our few belongings, and my father requisitioned living quarters for us in yet another house, using another war permit.

The council house we were billeted in was owned by a Mrs. Barge, who had a nine-year-old daughter and a five-year-old son. Mr. Barge was in the medical corps, stationed in Libya as part of the Eighth Army. For the next year and a half, I shared a room with Mrs. Barge's daughter, my parents had the main bedroom, while Mrs. Barge and her son slept in a third room.

The Germans bombed the port of Glasgow in 1941 and 1942, trying to destroy its shipbuilding industry, which was located about five miles from Mrs. Barge's house. Over a period of three nights, we suffered a serious blitz, as the bombing raids over Glasgow increased in intensity. The six of us had to sleep in an Anderson air raid shelter in the Barges' back garden. It was partially underground, and was made of corrugated iron. Most houses had these air raid shelters in their gardens during the war.

On the second night of the blitz, the Germans dropped incendiary bombs on our neighbourhood. They destroyed the church at the end of the street, and many houses near us caught fire, but we were spared. The next day, I walked around in the garden picking up pieces of shrapnel from the anti-aircraft shells. They had razor-sharp edges. One had to be careful going from the house to the shelter when the bombing started, because fragments of shells were dropping from the sky. One night when in a panic we were running from the house to the air raid shelter, my parents donned steel helmets and ran on ahead, leaving me to reach the shelter bare-headed, since all the helmets were taken. This scene continues to haunt me today, a reminder of the horrors of war suffered by children.

During the three days the blitz lasted, about 1500 civilians were killed in the port of Glasgow, the highly populated area of Clydebank and in our neighbourhood. Roughly 60,000 civilians were killed during the bombings of England and Scotland. This compared to the 400,000 casualties suffered by the British soldiers and air force and navy personnel. In addition, of course, were the many thousands of severely wounded in the civilian population and the military, and the more than one million houses destroyed or damaged in the air raids on London alone.

Soon it was time to move again. In late 1943, my father, now a captain in the intelligence corps, was posted to the east coast English port of Grimsby and Hull, where he was put in charge of security, checking the fishing boats that came in from Denmark, again potentially containing Nazi spies. This time my father succeeded in getting me accepted by Hymer's College, even though my scholarly aptitude did not impress the headmaster. At the age of eleven, I was a day student at this prestigious English public school, while living in a boarding house in Hull with my parents and twelve other people.

*

The war ended in 1945, and my father, now a major, was sent to northern Germany the next year to oversee the Flensburg occupation troops. Instead of following him once again, my mother and I returned to Copenhagen by boat in January 1947, and I entered a Danish high school. Since by this time, after all our years in England,

I was neither speaking nor writing Danish, the first year of school in Copenhagen turned out to be another gap in my education. It was a strange experience, sitting in classes not understanding a word that was being said around me, particularly by my teachers. I spent a lot of time looking out the window of the schoolroom, daydreaming.

In Denmark, children finish high school in their mid-teens and are then considered for a preparatory program for university in the gymnasium. Entrance to the gymnasium stream in Denmark depended upon tests, academic performance during high school and, most important, upon the recommendation of a gymnasium teacher. When I was fifteen, in my last year of high school, and by this time fluent in Danish, I was interviewed by a young mathematics teacher from the Copenhagen gymnasium. My whole future seemed to be riding on this interview. If the teacher was satisfied with my answers to his questions, he would recommend that I enter the gymnasium in the fall. Three or four years later, I could then apply to university. If I failed this personal interview, the only options available to me after high school would be entering a trade school or seeking work, most likely menial and ill paid.

The teacher ushered me into a small schoolroom with grey-painted walls and wooden desks. He cleared his throat, wrote on the blackboard and asked his first mathematical question. I sifted through the confused jumble of mathematics knowledge in my mind—detritus from my thirteen schools and two languages. When I did not immediately answer the teacher, he sighed and tossed another question at me.

Standing helplessly at the blackboard with the teacher's increasingly unfriendly blue-eyed gaze boring into me, I sensed the familiar feeling of a panic attack beginning. My heart raced, I felt faint and my brain ceased to function. The teacher, his voice rising, asked two more questions. I was unable to answer even the simplest one. I had become mute.

The teacher strode up to the blackboard, snatched the chalk from my hand and said, "Moffat, I can guarantee that you will never become a mathematician." I stumbled from the room. The interview had taken half an hour. The teacher's report was so negative that there was no chance at all for me to enter the gymnasium. That was effectively the end of my schooling.

*

Millions of children grew up during the Second World War in Europe. I don't know how many of them had as disruptive a childhood as I had, moving from town to town and school to school sometimes several times each year. I don't know how many others were subjected to such frightening and frequent bombing attacks and other horrors of war that they sustained permanent psychological damage. Millions of children today, in various parts of the world, still experience daily conditions of war, terrorism, natural disasters and famine. Once one gets beyond a sense of gratitude at having survived at all, it is natural to wonder how a wartime childhood helped shape the adult person one has become.

My inferior, fragmented schooling, together with the panic attacks induced by many nights of fearing imminent death by German bombs, seemed to prevent me from pursuing a life in science—or any other academic field I might have chosen. Yet it may also be possible that the conditions of my childhood helped me to develop the self-reliance and strong personal motivation necessary to overcome those limitations. If I had managed to survive the war, all the dangerous travelling my parents and I had to do, the fierce bombings, the lack of opportunities to form long-term friendships, the constant moving from school to school, then surely I could continue to survive and make something of my life.

2

————

TWO PATHS DIVERGED

A FTER my humiliating interview with the gymnasium teacher, and after graduating from high school, I considered seriously what I should do with my life. If my fate was not to earn a university degree, I mused, perhaps I should return to my early talents as an artist.

I had begun painting in earnest, outside the school environment, when I was fourteen, joining my father at his easel. While my father concentrated on abstract paintings, I painted abstracts and landscapes as well. Now, at sixteen, and unsure of how to proceed in my life without the advantages of a gymnasium education, I was again painting alongside my father in the evenings. This was satisfying to us both, and I believe it helped my father improve his outlook on life, as he was struggling to recover from tuberculosis.

After leaving the military at the end of the war, my father had started an import-export business in Copenhagen with a businessman who had been a prisoner in a concentration camp in Germany during the war. The business failed, and my father found a job in a larger import-export business in Copenhagen. It was while he was working there that he contracted tuberculosis. One of the female employees, a German national who had moved to Denmark, had a severe case of TB as a result of the deprivations she'd suffered during the war. Unfortunately, the employees all shared the same coffee cups during breaks, and my father contracted TB, becoming an invalid for a year or more. This created serious financial problems for my parents. My mother was forced to work as a waitress in restaurants in Copenhagen, and I contributed to the household income by working days on any odd jobs I could find, such as being a messenger boy and washing windows in the apartment blocks in our neighbourhood in Valby, a suburb of Copenhagen. I also worked for several months as delivery boy for a florist in Copenhagen. My job was to go to churches prior to funerals and place a bouquet of flowers on the corpse lying in an open coffin. This evoked in me a dread of death as powerful as any of my experiences and fears during the wartime bombing raids, as I stood in the quiet church contemplating someone's dead relative and the fleeting years of life.

When I was fifteen, I saw an exhibition of paintings by the Russian-French abstract painter Serge Poliakoff in a small gallery in Copenhagen, and was very impressed with his work. I was struck by the marvellous juxtaposition of brown, red and blue in his paintings, as well as the unique combination of abstract forms. I began to daydream of going to Paris to become an artist, taking lessons from Poliakoff if he would have me. Working at my odd jobs since leaving school, I managed to save enough money to journey by train to Paris. In contrast to my father's experience, in which his father forbade him to go to Rome on an art scholarship, my father strongly encouraged me to pursue this dream.

In my youthful ambition, I imagined my work being exhibited in major galleries in Europe and America. I was excited by the idea of living in Paris and becoming part of the bohemian artistic life there. I had developed a deep passion for my art, and could imagine no other life than the pursuit of beauty through painting.

*

I had just turned seventeen when I boarded the Northern Express from Copenhagen to Paris in 1949. We stopped in Hamburg for two hours, and I walked around the Bahnhof, the railway station. It was only four years since the war had ended, and I was astonished to be able to look from the Bahnhof out to the horizon of the city, because so many of the major buildings in Hamburg had been flattened by the unremitting Allied bombing. I also saw several German veterans in grey uniforms, missing arms or legs, walking around with the help of crutches and begging for money.

When I arrived at Gare du Nord in Paris early the next morning, I collected my bicycle from the luggage car and hoisted my satchel containing all my belongings onto my back. I decided to do a little sightseeing on my first day in Paris, and took a detour through the expansive Place de la Concorde on my way to the Seine. It was a bright, sunny spring morning and due to the shortage of gas, there were few cars on the roads. Travelling on Parisian streets by bicycle then was a safe venture, in contrast to today.

All my expectations of what Paris would be like were met on that morning. The tall, grey buildings with their iron-grille balconies, colourful potted plants and dark grey slate roofs with innumerable chimney pots formed a backdrop for the intense activity of the city. I bicycled past the many cafés where waiters in white aprons were moving chairs and tables out onto the pavement. Street cleaners were flushing water down the gutters, and there was a distinct smell of Paris that I can still recall: a clean smell sharpened by the musty odours of an ancient city.

I crossed Pont Saint-Michel and stopped halfway across the bridge to watch the grey-green, muddy water of the Seine passing below, and the barges docked at the stone mooring walls. Looking up to my left, I was overwhelmed by the soaring towers and huge rose window of Notre Dame, and the gargoyles springing out of the stone walls that shimmered in the early-morning light. Turning the other way, I saw in the far distance the massive edifice of the Louvre.

I bicycled on up Boulevard Saint-Michel and arrived at the Place Denfert-Rochereau with its massive stone lion gazing mournfully over the Parisian scene. I was on my way to Porte d'Orléans on the outskirts of Paris where I had arranged by letter to rent a room.

After settling into my accommodations, I soon discovered rue de Seine, the part of town where most of the famous art galleries were located, and luckily, one was showing Poliakoff paintings. I asked the gallery owner how I might contact Poliakoff, and he gave me an address on rue Madame, on the Left Bank, where Poliakoff had a small studio and also lived with his wife and child.

One afternoon I plucked up my courage, located the artist's street address, entered an inner courtyard and knocked on the door of a little room that was behind a toy shop facing rue Madame. I was feeling apprehensive as I waited for someone to answer my knock. Would today be a new beginning in my life, or was I to be disappointed again?

A striking-looking man with black hair and lively brown eyes opened the door. He was friendly to me, an unexpected visitor, and introduced me to his wife, an Irish woman whom he had married while studying at the Slade School of Fine Art in London. Poliakoff spoke fluent English as well as French, so we conversed in English. After I explained to him how I had admired his paintings in Copenhagen and had come to Paris recently to pursue an art career, he invited me to spend time with him in his studio, where I became his student.

Serge Poliakoff was very kind to me, accepting me as his only student even though I had no money to pay him, and spending a year teaching me the techniques of abstract painting. He taught me, in his unique way, how to make oil paints from the original powder sold at a special shop in Paris. This technique of producing oil paint was one of the secrets of the vibrant colours in his paintings. He also told me to visit the Louvre once a week in order to appreciate and learn from the works of the great masters. At this time, Poliakoff was living in quite poor circumstances. As painting was not providing a living, he also played the guitar in cafés at night, specializing in Russian folk music.

My small rented room in Porte d'Orléans was in a turn-of-the-century greystone apartment building in rue des Plantes. I would paint my abstract canvases by propping them up against an ancient wooden wardrobe opposite the large oak bed, making sure that I didn't drip too much paint on the Turkish rug, which would upset my landlady, who took a motherly interest in my welfare. My paintings were in the style of abstract expressionism developed by the Parisian school in the 1940s and '50s. I used bold colours and developed planes of colour with expressive brushstrokes.

One day, Poliakoff told me to bring five of my paintings to his studio. We would each hang five of our paintings in a spring art show—the Salon des Réalités Nouvelles—an annual event held at the Musée d'Art Moderne on avenue du Président Wilson. Our paintings hung among those of artists such as Victor Vasarely, Pierre Soulages, Jean Dewasne, Hans Hartung, Jean Deyrolle and other abstract expressionists. Poliakoff himself became famous as an abstract painter before he died in 1966, lifting himself and his family far out of the impoverished circumstances I had observed when working with him in rue Madame. Today his paintings are auctioned at high prices and exhibited in major art museums worldwide.

During my year in Paris, my father came to visit for several days. He took the train from Copenhagen and stayed in a small hotel near my rented room. I had the uncomfortable feeling that he was living vicariously through me. My father was still painting at that time and was eager to meet Serge Poliakoff. Perhaps he envied me being able to enter the world of art in Paris as a young man, for his chances of doing so in Rome had been thwarted by his father.

Although in reviews of the spring Salon show in *Le Monde* and *Figaro* I was flatteringly described as a young, upcoming art talent, after a year in Paris I ran into severe financial difficulties, as my meagre savings were running out. During the immediate postwar years, employment opportunities were scarce in Paris, particularly for foreigners. To help my financial position, I sought out American tourists, offering to be their guide. After showing them around places of interest, I would meet them at their hotel or in a café, a painting or two in hand, and attempt to make a sale. Inevitably these guided tours were financially unsuccessful, although I generally managed to obtain a free lunch or dinner.

Reluctantly, after my year as an aspiring painter, I returned home to Copenhagen. I was discouraged by my failure to begin a successful art career in Paris, and I knew that I would not have such a year of opportunity again. I did manage, however, to mount a show of Serge Poliakoff's work and mine at Illums Bolighus gallery that year in Copenhagen. Poliakoff did not come in person, but sent ten canvases to me by train. We did not sell a single painting, and I felt even more disappointed about my lack of progress in the art world. Copenhagen, after all, was not a centre for art like Paris, and I knew that I would be inviting continuous financial difficulties if I devoted myself to painting.

Again I found myself living with my parents, my father still not fully recovered from tuberculosis and my mother working long hours in downtown restaurants. Just as before my year in Paris, I again worked every day at boring or distasteful odd jobs to bring in money to help support us.

And again, I was forced to ponder what to do with my life.

*

Not long after returning from Paris, out of curiosity I picked up two popular-science books, *The Nature of the Physical World* and *Spacetime and Gravitation*, both written by Sir Arthur Eddington. They were about Einstein's theory of relativity, cosmology and the evolution of stars. These books greatly affected me. It's as if they turned a switch on in my brain that I had no idea was even there. Eddington, a fine writer, was able to create a sense of almost spiritual wonder in me at the mysteries of the universe and an emotional desire to know the truth of how the universe began. His books triggered an astonishing turning point in my life.

After reading the books, I began having strange visions of the structure of the universe and the fabric of spacetime as revealed by Albert Einstein. In these daydreams, I tried to comprehend how the universe was structured. These daydreams were intuitive forms rising from my subconscious rather than conscious attempts to understand the universe. The visions seemed to indicate some primal urge developing in me to connect with the stars and galaxies of the universe.

Initially my visions were colourless, and then they turned into vast, colourful canvases. I began to realize that there was an unconscious merging of my visual experiences when painting and the visualization of the heavens all around us. When I painted, I didn't contemplate the "meaning" of art or feel any phenomenological need to "prove" my paintings. But as I continued to read and daydream, I began to realize that physicists who attempt to understand nature initially have a visual experience which then has to be transformed into a theory by means of mathematical formulations. However, in contrast to creating a painting, this initial imaginative process in physics has to be verified eventually by experiment. Later in life I expressed this idea as: Physics is imagination in a straitjacket.

I decided, against the incredible odds, that I would try to pursue science seriously, particularly physics and mathematics. Obviously, I could not contemplate becoming an experimentalist, for this required special academic training and access to laboratories and experimental apparatus. However, I could pursue theoretical physics with just my brain, pencil and paper, and access to a good library. I tried to push aside my knowledge gaps and failures as a student. At none of the many schools I attended had I thrived intellectually. One principal had counselled my parents to put me in trade school. I tried to push aside the memory of the mathematics teacher whose judgment had killed my chances for admission to university. Since that normal route for acquiring knowledge was closed to me, I would have to pursue this new dream on my own. I was at a stage in my life when one's ardent desires and passions can overcome what to others would be impossible odds. I had a self-confidence then that is perhaps only possible when one is nineteen.

In counterpoint to the memories of my failures, another strange memory rose in my consciousness. I recalled that when I was six or seven, my father took me to see a psychiatrist in London because I insisted on reading the time in counter-clockwise fashion, and I also reversed whole sentences when learning to read. The psychiatrist peered at me with curiosity, and asked me questions in staccato sentences. When he was finished with me, I sat in the waiting room and overheard him telling my father that his boy was a "genius." At the time, I turned the word "genius" over and over in my mind, and decided that it could not mean anything favourable for my future.

But now I thought about it again. I discovered during that year in Copenhagen, and contrary to all my experiences in school, that I had a surprising, indeed remarkable, ability to learn mathematics and physics rapidly. This was partly due to my photographic memory, which I also first discovered during that year. Now I wondered whether my talent for learning science and mathematics rapidly had been there all along, untapped by any of my teachers, or whether it might actually have developed suddenly, perhaps even as a consequence of the post-traumatic stress disorder I had suffered after the war. Despite all my previous failures, I was now highly motivated to excel in mathematics and physics; perhaps it was my strong motivation itself that released my aptitude for science and math, which had been dormant all those years. And my motivation sprang from my intense reaction to the Eddington books. Could there be a more compelling testimonial for good popular-science writing than this?

It was fortunate for me that the University of Copenhagen library allowed people to borrow science books and periodicals without their having to be enrolled as students. In this way I was able to read physics and mathematics materials and move quickly towards an understanding of modern physics and cosmology. I taught myself stealthily, sneaking time to go to the library while on my messenger jobs, and poring over my books and papers every night. My parents didn't know what to make of this latest development. Our financial situation was worsening, and they feared that this new turn in my life would distract me from contributing to the support of the family. Yet I did manage to continue working at my menial jobs, while at the same time completing the equivalent of four years of undergraduate training in physics and mathematics within the year. I learned the basics of calculus in less than two weeks and became proficient in solving differential equations.

*

Because I had progressed rapidly in my private studies, I began feeling that surely, with my newly discovered aptitude, I could do something more interesting than the odd jobs that provided me with a meagre wage. I decided to push things further. I made an appointment with the director of the astronomical observatory in

Copenhagen. I explained to him that I was interested in becoming a physicist, and was studying physics, astronomy and cosmology by myself. He sent me to the Copenhagen University Geophysics Institute, where the kindly director took an interest in me, and gave me a job solving least squares calculations of gravitational measurements. These calculations went into large published tables of gravitational measurement data and were used in geological surveys as well as in searches for minerals and other resources in Greenland, which was part of Denmark.

Now at least I had an occupation in a scientific field! I couldn't help thinking about Einstein's somewhat similar position as a Swiss civil servant in the patent office in Bern, where in the miraculous year of 1905 he published five papers in his "spare time," including one on the special theory of relativity, which revolutionized physics.

After my first intense learning period, I began concentrating on relativity theory and advanced through Einstein's special relativity and general relativity theories to his most recent work on unified field theory, which was his attempt to unify gravity and electro-magnetism within a geometrical spacetime structure. Einstein had the brilliant idea that the force of gravity, as first envisioned by Isaac Newton, was not actually a force of attraction between two massive bodies. Rather, it was the effect of one massive body distorting or curving the spacetime geometry around it, which in turn affects another body nearby. Einstein's idea that spacetime geometry is curved in the presence of matter was the basic tenet of his general theory of relativity.

In Einstein's special theory of relativity, he envisioned velocity as "relative." That is, an observer moving in a non-accelerating frame of reference cannot tell whether he is at rest or moving, or how fast he is moving, in relation to an object in another non-accelerating frame. This concept was easy for me to grasp, thinking of the momentary confusion of sitting in one moving train and watching another that is also moving.

In his general theory of relativity, Einstein envisioned acceleration—or changing velocity—as also "relative." In fact, he proposed the startling idea that gravity and acceleration are equivalent. Picture a skydiver falling before she opens her parachute. With her eyes closed, she would not be able to tell whether she was falling due to the pull of the earth's gravity or to a force exerting an acceleration upon her.

Relativity was an idea that was in the air at the beginning of the twentieth century. Others besides Einstein, such as the mathematical physicists Hendrik Lorentz and Henri Poincaré, had formulated theories of relativity beyond those already envisioned by Galileo in the sixteenth century. However, they could not free themselves from the concept of the "ether," the supposed substance that permeated all of space and allowed electromagnetic waves to travel through it, which virtually all scientists believed in at the time. It took the genius of Einstein to ignore the concept of the undetected ether, to make special relativity a universal property of space and time, and to develop a classical mechanics that was compatible with special relativity.

Relativity, however, constituted only half of Einstein's effort to create a unified theory. There is also the concept of "fields." In the nineteenth century, James Clerk Maxwell's equations unified the electric and magnetic fields. These fields were first conceived by Michael Faraday, who pictured them as lines of force originating from electrically charged particles or magnets. These fields can be observed in the regular lines formed by pieces of metal filings on a sheet of paper when it is held above a magnet. In Maxwell's theory, the electromagnetic fields exist in four-dimensional spacetime, which acts like an arena in which the fields themselves and electrically charged particles move like hockey players in an ice rink. Einstein wanted to unify his geometrical theory of gravity with

Maxwell's equations for the electromagnetic fields into one unified theory. In 1918, the famous German mathematical physicist Hermann Weyl had proposed a way of unifying Maxwell's theory and Einstein's gravity theory that was not successful. In his later years, Einstein continued, also unsuccessfully, to try to find better ways of unifying gravity and electromagnetism.

I quickly became caught up in this quest for a unified field theory, and studied Einstein's papers closely. I had been checking through the basic calculations underlying Einstein's latest unified field theory, and I discovered that one of its basic assumptions had what I considered a flaw. I composed my first physics paper on this subject.*

After composing this paper, feeling intense excitement as I wrote it and calm satisfaction when I reviewed it, I began to dare to think that yes, possibly I could have a career as a physicist, and began considering steps that I could take to achieve this. At age nineteen, I should have been in my second year at university, but I hoped to somehow find another route to this new goal.

*

My father had made the acquaintance of an American chemist who was conducting research in the laboratory of the Carlsberg brewery in Valby, not far from my parents' apartment. This gentleman expressed an interest in meeting me, as he thought he might be able to help me achieve my goal. He was acquainted with John Page, an assistant to the British consulate in Copenhagen.

My father agreed with his friend that I should go through the British consulate for help because I was still a British citizen.

---

*In technical terms, I was questioning the need for Einstein's action principle based on a nonsymmetric metric field to satisfy a real Hermitian symmetry. He had been hunting for the most satisfactory action principle that was the basis for the derivation of his unified field theory equations.

Although I had been born in Copenhagen, I was not considered a Danish citizen because my father was British. With my father's help, I composed a letter to Mr. Page, and explained that I was a nineteen-year-old student who had, through private studies at the university library, achieved enough knowledge of mathematics and physics to study Einstein's work on unified field theory, and had written a manuscript on his theory. I was now hoping to somehow enter the academic world and pursue physics studies, possibly in England. My father added a note explaining that he had been a major in the British army and had been stationed in Flensburg, Germany, at the end of the war as the district station commander. He thought that perhaps this would help establish our family's credibility.

About a week later we received a letter from Mr. Page showing much interest in my situation. He said that he had contacted the Niels Bohr Institute in Copenhagen and had spoken with Niels Bohr himself. Bohr, he wrote, wished to speak with me. The following week, I received a letter from Bohr's secretary setting up an appointment.

Within one extraordinary year in my young life, I had left the path of Serge Poliakoff and my aspirations to become an abstract painter, and was now starting down a completely different road, towards a door being opened to me by the greatest living Danish physicist, Niels Bohr.

3

———

NIELS BOHR

O N   T H E   M O R N I N G of my appointment with Niels Bohr,
I took a tram to Blegdamsvej 17 in Copenhagen, carrying
my manuscripts in a large brown envelope. By this time,
I had written a second manuscript about unified field theory. I
paused for a few moments outside the three-storey building with
its red-tiled roof and little courtyard facing the street, and gazed at
the tarnished brass letters spelling out "Niels Bohr Institutet 1920."
I suddenly felt my heart palpitating. What was I going to say? How
should I address Niels Bohr? Could he really help me?

Bohr had mapped out the structure of the atom in 1913, and
later in the 1920s had helped to develop the quantum revolution.
After Ernest Rutherford discovered that there was a hard core in the
centre of atoms, consisting of a positively charged nucleus, Bohr
produced a model of the atom in which the energy associated with
the spectral line radiation* was quantized, or occurred in discrete

---

*Atoms emit photons when an electron makes a transition from a particular dis-
crete energy level to a higher one, and they absorb photons when changing to a
lower energy state. The emissions or absorptions occur as coloured or dark lines,
respectively, within the radiation spectrum. These spectral lines are highly specific
to different atoms, and can be used to identify elements, for example, in the composition of stars.

quantum units. This followed the important discovery by Max Planck in 1900 that radiation emitted by a hot body was not continuous, as had been assumed in classical physics, but came in discrete parcels of energy. In effect, Bohr succeeded in making the atom stable. Previously, the classical model of the atom could not prevent the orbiting electron from spiralling in towards the nucleus. Bohr pictured his atom as a mini-solar system, with the positively charged nucleus playing the role of the sun and the electrons swirling in stable, planet-like orbits around it. However, Bohr's young assistants, Werner Heisenberg and Wolfgang Pauli, were not able to apply his model of the atom to more complicated spectral line data.

Eventually physicists gave up on the Bohr model as a true visual picture of the atom and developed the modern version of quantum mechanics, which described the electrons and protons in the atom as waves. This new view interpreted the emission of the spectral lines from the atom in terms of probability theory. That is, light, consisting of photons, had only a certain "chance" of being emitted from the atom at any particular time. This new description of the atom gave birth to what is now called quantum mechanics, which revolutionized physics by overthrowing the whole notion of classical physics applying at the subatomic level.

Because of his pioneering work in atomic physics, by 1953, when I met him, Niels Bohr was certainly the most famous scientist in Denmark, and was one of the most famous physicists in the world. He had won the Nobel Prize in physics in 1922.

*

I opened the gate, entered the courtyard and rang the bell next to a door that said "Administration." A buzzing noise answered me. I pushed the door open and climbed the stairs to an office where a secretary sat at a large oak desk overlooking the gardens and park at the back of the institute building. It was a clear spring day and through the window I could see a soccer game in progress in the park, with young men kicking a ball around the field. Through the glass I could hear their muted shouting and birds in the nearby trees as well. The secretary smiled at me and asked in Danish what my business was. I explained that I had an appointment with Professor Bohr at ten-thirty. "Ah, yes, you're the young man who is working on Einstein's unified field theory," she said, fixing me with a curious gaze. "You may sit in that chair over there and wait for Professor Bohr."

As I waited, my nervousness increased. I could feel my palms becoming moist. I tried to concentrate on the shouts of the soccer players, and block out all other thoughts. After several minutes a short, stocky, middle-aged man with curly black hair opened the office door and walked over to the secretary, ignoring me. She said, "Dr. Rosenkrantz, this young man has an appointment with Professor Bohr."

The man turned and looked at me for the first time. "Ah, you're the young man who is working on Einstein's unified field theory," he said.

"Yes," I replied. "My name is John Moffat." I stood up and we shook hands.

"Do you have your manuscript on the unified field theory with you?" he asked.

"Yes, I do," I said, holding up my large brown envelope, which shook slightly in my trembling hands.

"Well, then," he said, "let's go in and see Professor Bohr."

*

We walked down a long dark corridor and came to an office. When Rosenkrantz opened the door, the strong smell of tobacco smoke wafted out. A tall man with a dome-shaped head and wearing a rumpled brown suit stood in the far corner of the room looking out the window at the soccer game in the park. His hair was thinning, accentuating his large ears.

Rosenkrantz beckoned me into the room and performed the introductions. "Professor Bohr, this young man is John Moffat."

Niels Bohr removed a pipe from between his thick lips and said in softly spoken Danish, "So, you're John Moffat?"

I said yes, I was.

"Do you have your manuscript on Einstein's unified field theory with you?"

"Yes," I replied. Again, I lifted the envelope. "I have two manuscripts, Professor Bohr. One has to do directly with Einstein's classical unified field theory. The other attempts to quantize his unified field theory."*

Bohr then crossed the room to his desk, sat down and contemplated a wooden rack holding pipes of various sizes, standing next to a large Danish matchbox and several tins of tobacco. He took his pipe out of his mouth and, pointing the stem at me, said, in English this time, "Please, sit down, sit down." I took one of the chairs in front of his desk. Dr. Rosenkrantz sat in a chair next to mine, and placed a notebook on his knee. He took a pen out of his breast pocket, unscrewed the cap and began writing scratchily. I realized that he must be about to take notes, and my palms started sweating again. Was he actually going to record everything that was said?

There was a silence while Bohr picked up one of the tobacco tins from his desk, removed the lid, and poked his bear-like hand inside to pull out a lump of tobacco. He removed his pipe from his mouth and shoved the tobacco into the bowl. Then he carefully took a

---

*By "quantizing," I meant that the energy associated with a field such as the electromagnetic or gravitational field must also come in quantum packages of energy. This is accomplished by "quantizing" the classical field theory through the application of the mathematics of quantum mechanics and relativity theory.

metal instrument from his pipe stand, tamped the tobacco down into the bowl with it and put the pipe back into his mouth. Next, he picked up an oversized box of matches, took out a match and lit the pipe, making dry sucking sounds as he drew in the smoke. Printed on the matchbox was the familiar Danish trademark H.E. Gosh & Co., with the famous picture of Tordenskjold, the eminent Danish-Norwegian naval hero of the seventeenth century. Everyone had these matchboxes in their kitchens. Bohr sat back in his chair and studied me for a while. The room was silent except for the birds singing outside. The shouting of the soccer players had ceased. Bohr said, "Let's see the manuscript on unified field theory."

I opened my envelope, took out the manuscript and leaned across the desk to hand it to him. I had managed to make copies of the two papers on my parents' old Royal typewriter with faded carbon paper, and had written in the equations with a pen. Bohr put the pipe down on what looked to me like an antique Royal Danish porcelain dish, and spent several minutes quietly reading my paper. He then said, "Hmm!" and looked up and smiled at me.

While he had been reading, his pipe had gone out. He took another one from the stand and followed the same routine again: filling the pipe, tamping down the tobacco, lighting it and sucking it into life. Then he leaned back in his chair staring at the ceiling while the blue smoke rose up from his pipe and filled the room with its strong, sweet aroma. Thus we sat for at least five minutes waiting, while Bohr thought about what he had read. Rosenkrantz sat quietly with his fountain pen poised over his notebook.

Finally, Bohr levelled his gaze at me and ruminated, "So, you've been working on Albert's unified field theory."

"Yes," I said.

Bohr took the pipe out of his mouth, as apparently it had gone out again. It joined the first pipe on the porcelain dish. I wondered if he was going to light another pipe before we proceeded with the business at hand. And indeed he did. He removed a third pipe from the stand and performed the same ritual yet again, while I counted up the pipes still in the stand. I thought that if we were going to go through those remaining ten pipes, I would be in Professor Bohr's office for the rest of the day.

After he succeeded in lighting the third pipe, he asked me in a mumbling voice, "What is your opinion about Einstein's efforts to unify electromagnetism and gravity?"

Now Dr. Rosenkrantz's pen began scratching rapidly in the notebook. I sensed that something serious was finally about to happen in my meeting with Niels Bohr.

"I think it's a logical extension of his gravity theory," I replied.

Bohr smiled, removed the pipe from his mouth and said, "Are you aware of the fact that Albert has become an alchemist?"

I was taken aback by this obvious slight of one of the great physicists of the twentieth century. I said, "You mean someone who tries to turn base metals into gold?"

Bohr smiled again, and mumbled, "Yes, something like that." Bohr's derogatory comment indicated that he felt strongly that Einstein was wrong in denying the probabilistic nature of quantum mechanics. Einstein did not believe that quantum mechanics formed a complete description of reality, whereas Bohr did. Instead of Einstein working to develop quantum mechanics further, he had been focusing on extending his theory of gravity to include the electromagnetic forces, but without including the nuclear forces, which in the opinions of contemporary physicists, played a more significant role than gravity in the subatomic realm.

I had to concentrate on catching his words because Bohr mumbled even more disturbingly than most Danes do when they speak, whether in Danish or English. Evidently Dr. Rosenkrantz, who was also Danish, was used to Bohr's inaudible mumbling and was able to follow his conversation more easily than I.

I sat for a few moments, wondering how to respond to this rebuke of Einstein. "I understand, Professor Bohr, that you do not believe that Einstein is following a satisfactory path in his physics." Miraculously, my initial nervousness had disappeared, and I was now entirely focused on the conversation and on figuring out Bohr's attitude towards Einstein.

Bohr put down the pipe, which had now gone out, took another one from the rack and began the ritual. I now understood why he had such a large box of matches. When he had finished lighting the pipe, he said, with the pipe stem still in his mouth, "I feel that Albert has been wasting his time. You cannot ignore quantum mechanics and hope to achieve any success by unifying the classical gravitational and electromagnetic fields, in the way he attempts to do this."

I said, "Well, Professor Schrödinger has also been attempting to unify gravity and electromagnetism, using the same formalism as Einstein, the nonsymmetric theory."

Bohr removed the pipe from his mouth, looked at me sternly and said, "Well, Erwin is also an alchemist these days. He is pursuing this foolish denial of the successful development of quantum mechanics, and has also gone down this blind alley of trying to unify the classical gravitational and electromagnetic fields."

Then I asked, "So, Professor Bohr, you believe they should be following a quantum mechanical interpretation of these fields?"

He then mumbled, "We already have a quantum mechanical interpretation of electromagnetic fields, which has been quite successful. As for gravity, I do not see that a classical unified theory of gravity and electromagnetism can ever solve the problems of atomic physics."

"I understand," I said, "that Professor Einstein does not agree with your interpretation of quantum mechanics. In fact, he seems to be opposed to quantizing the gravitational field."

"Well," Bohr said, "it saddens me to see that Albert has taken the wrong path in recent years. We have disagreed about this issue for years, and he refuses to accept the success of the quantum theory. He insists on being pigheaded about this."

We sat in silence for a while again, and I waited for Bohr to put down his pipe and start over again with another. And indeed he did. I wondered if he went through this ritual every time he granted an interview. When he succeeded in lighting the fifth pipe, he leaned back in his chair and looked at me. "Anyway," he said, "how did you come to write this manuscript? I understand from Mr. Page at the British consulate that you've been studying mathematics and physics by yourself."

I nodded, and shifted in my seat. Now that the conversation had turned from physics to me, I began feeling nervous again.

"When did you leave school?"

"I finished high school in Kaptain Johnson's school in Copenhagen when I was sixteen and then pursued an art career in Paris."

"So you didn't go to a gymnasium?" Bohr asked.

"No," I said.

"And you haven't attended the university in any way?"

"No," I said again.

"So where did you get the books to learn mathematics and physics?"

"At the university library. The science textbook and periodicals section of the university library is open to the public."

I saw Bohr and Rosenkrantz exchange glances.

"How long did it take you to learn mathematics and physics, such that you could write this manuscript?"

"About a year," I answered.

"By yourself?" he asked incredulously.

"Yes."

"So what do you plan to do now?"

"I'd hoped that I could pursue an academic career and study physics, possibly in England. I am a British citizen."

"Well," said Bohr, removing his pipe from his mouth, "you couldn't pursue an academic career in Denmark without going through the usual channels. That is, you have to take your student exam at a gymnasium and then enter the physics curriculum at the University of Copenhagen. Would you want to do this?"

"I think that I would prefer to go to England if possible, to an English university."

"Well, then," he said, "you would find that you would have to proceed through the same kind of academic channels that you would here, but it is possible that it would be easier for you than here in Denmark."

I then plucked up my courage, and said, "Professor Bohr, would it be possible for me to give a talk here at the institute? I have been working on some quantum mechanical interpretations of gravity, using Julian Schwinger's approach to quantum field theory."

As we were talking, I had suddenly had the rash thought that if I could give a convincing talk to the physicists at the institute, they might accept me as a promising young physicist and I could perhaps bypass those "usual channels" that Bohr had referred to. During my private studies of physics, I had gone beyond quantum mechanics and learned about quantum field theory, combining quantum mechanics and the special theory of relativity.

Again, Bohr exchanged a glance with Rosenkrantz, who was still busy recording all that we were saying.

"Do you think you are capable of giving such a talk here at the institute?" Bohr asked.

I shifted nervously in my seat and replied brashly, "Yes, I think I am, Professor Bohr."

"Will you be using Einstein's unified field theory?" he asked.

"Yes, I think I will, as well as some ideas I have about quantizing his theory," I responded.

"I'm afraid that you will meet much skepticism," Bohr said. "Professor Christian Møller, who is our expert here at the institute on gravity, is also skeptical about Einstein's program, but we can arrange for you to speak to him and see if he can arrange for you to give a seminar."

Bohr put down his pipe, to join the others in the porcelain dish, and directed Rosenkrantz to arrange a meeting with Møller.

"Also," Bohr said, addressing me again, "I will speak to Mr. Page at the British consulate and see whether they can help you with possibly pursuing your studies in England."

Bohr stood up, signalling the end of the interview, and Rosenkrantz and I also rose from our chairs. I shook Professor Bohr's hand and thanked him for his kindness in having this meeting with me. Rosenkrantz and I left the smoke-filled room together.

4

ALBERT EINSTEIN

Two weeks later, I gave my talk in the seminar room on the ground floor of the Niels Bohr Institute, with its pleasant view of the adjoining park. The audience, consisting of Professor Møller and other professors and their students, was hostile, just as Professor Bohr had predicted. They made rude comments to me about Einstein's unified field theory research and my attempts to defend it. I felt disappointed that Professor Bohr himself had not attended my talk.

Afterwards, Professor Rudolf Haag, who was visiting the institute from Germany, came up to speak to me. He advised me to enter the university and go through a conventional academic training to obtain an undergraduate degree. He agreed that it would be best to do this in a British university. Haag was the one person at the institute that day who showed any interest in my research and my future.

Disturbed by my overall reception at the seminar, I returned home, and with the rashness of youth, I wrote a letter by hand to Albert Einstein, describing the negative reaction of the audience towards his goal of unifying gravity and electromagnetism in one geometrical scheme. I also included the two manuscripts that I had written on his unified field theory. ". . . I have today held a talk on my work with regard to your theory at the Niels Bohr Institute, Copenhagen," I wrote, "and there found only complete misunderstanding. In fact, it appeared that the main purpose was to undermine my personal confidence as to my ability. I would be eternally indebted if you could find time to read my work, and should you find yourself satisfied with my interpretation, mathematical ability, and conceptions, that you return same to me with your opinion . . . I found the Personel [*sic*] at the Bohr Institute completely without the fundamental knowledge necessary for the understanding of your Theory of Generalized Gravitation, and need I state how great a disappointment this was to me.

"Moreover," I continued helpfully, "I found the attitude of the persons concerned contrary to all aesthetic feeling and conceptions, and I feel that you should become aware of what is really going in the opposite camp."

I posted the letter, along with my two manuscripts, one of which pointed out a potential problem in his unified field theory, to Einstein at the Institute for Advanced Study at Princeton University, New Jersey. My manuscripts were titled "Theory of Quantized Unified Fields" and "On Unified Field Theory and the Equations of Motion."

The year was 1953. I never expected a response. It had been a valuable learning experience for me just to write the papers and to give the seminar. Why would I anticipate that Einstein would have time, great physicist that he was, to read and respond to my letter and manuscripts? Perhaps he would be offended that an unknown, unschooled young man in Copenhagen dared to criticize his published theory.

*

Three weeks later, an airmail letter with U.S. stamps arrived for me at my parents' apartment. The return address was Einstein's home at 112 Mercer Street, Princeton, New Jersey. Even then, in my excitement, I reflected on the fact that Einstein did not use the Institute for Advanced Study as his official address. I tore open the envelope and found that Einstein had written back to me in his own handwriting! He had also included a reprint of the appendix to his latest edition of *The Meaning of Relativity*, which I had not yet read. The book I had borrowed from the university library was an earlier edition.

But unfortunately, Einstein's handwritten letter was in German, and my German from two years of study in two different Danish schools was not up to the task of understanding it.

I immediately thought of my barber, Hans Busch, who was of German extraction. I rushed down the street on that warm June afternoon and burst into Hans's shop shouting, "I just received a letter from Albert Einstein in America! It's written in German. Can you please help me translate it?"

It was midmorning, a busy time in the small shop, and the men sitting and waiting for their haircuts stared at me in astonishment. The barber looked at me, his scissors clipping at a customer's hair, and said, "Just a minute, John. I need to finish with this gentleman."

I curbed my impatience and sat down to wait, feeling exhilarated.

Finally, Hans Busch's elderly client rose from the barber chair, brushing stray hairs off his jacket, and paid for his haircut. He joined the other customers seated around me to watch the show.

"All right, John," Hans said, wiping his hands on his apron and turning to me. "What's this you have here?"

I pulled the letter out of my pocket and handed it over. It was written on two sides of a piece of stationery. Hans perused it quickly and stopped at the end, peering at it intently. "Well, indeed, this letter has Albert Einstein's signature!" The other customers stood up and gathered round to stare at the signature too. The barber said, "What have you been up to, John?" and he smiled. I gave him, and our audience, a brief summary of how I had spent my spare time during the past year, the talk I'd given at the Niels Bohr Institute, and why I had written a letter to Einstein.

Hans proceeded to translate the letter into Danish orally, despite some difficulties with the technical terms in German, while we all stood listening attentively. According to the translation, Einstein made constructive comments and criticisms on my manuscripts, and attempted to answer my criticisms of his mathematical formulation. He also had some things to say about my audience at the institute.

"'I can understand very well that your work has not found a favourable reception in Bohr's circle,'" Hans translated. "'For every individual and every study circle has to retain its own way of thinking, if he does not want to get lost in the maze of possibilities. However, nobody is sure of having taken the right road, me the least.'"*

Einstein went on to make some profound comments about physics as it was at that time. Later, I was able to obtain an English translation of the letter, and had more time to contemplate Einstein's ideas. Indeed, his comments still have significance for many of the endeavours of physicists today:

I do not believe that one achieves one's objectives by first setting up a classical theory and then quantizing it. Although this, of course, has been successful in the interpretation of classical mechanics and in the interpretation of quantum facts by modifying this theory according to the principles of statistics. But I believe that in the attempts to carry over this method to field theories, one encounters ever increasing complications and the

---

*Quotations from Einstein's letters in this chapter are used with the kind permission of the Albert Einstein Archives at the Hebrew University of Jerusalem, Israel.

necessity to increase the number of independent assumptions monstrously. For generally covariant field theories* this will be even worse.

I even think that the mechanics based on the quantum theory cannot provide a useful starting point for a more profound theory, despite its significant successes, because I can see by looking at it that it has accepted the understanding of the "quantum jumps" in an "illegal" fashion by raising probability to a reality and in doing so giving up the reality of the quantum states (in the old sense). Thus one wants to explain why an apparent *arbitrarily small* perturbation can change the energy of an atomistic system by a *finite* amount [Einstein's italics].

I understood from Einstein's words that he believed that the old quantum physics that he had been involved with, in the early days of the birth of quantum mechanics, represented a complete description of reality, whereas the later quantum mechanics and quantum field theory, which were accepted in the 1950s, were based purely on probability theory and statistics, and did not constitute a complete and real description of nature.

Einstein continued:

In view of this state of affairs I see myself urged to consider the logical simplicity as the *sole* guide using general relativity. This leads me to the attempt (but not to the conviction) to seek the

---

*Einstein's general theory of relativity required that the field equations of the theory remain the same for all observers, such as those floating in space, sitting in an accelerating rocket or standing on the surface of the earth in the earth's gravitational field. This principle of covariance (or invariance) generalized the theory of special relativity, in which the laws of physics were only invariant with respect to inertial frames of reference, i.e., for observers who are not accelerating and are not subject to gravitational fields.

future in a field theory (in the old sense) (generalization of the theory of the gravitational field). The point of view that one is not allowed to construct the Lagrangian function from logically independent terms appears essential to me. I send you my latest research from which you can see what I mean by that.

I was overwhelmed by the amount of obviously pertinent information one of the world's greatest physicists had imparted to me, and I understood the significance of Einstein's comments for the future of physics. It is clear from his comments on quantum mechanics and the way it was applied at the time that he had not wavered in his opposition to Bohr's interpretation. The famous discourses on the meaning of quantum mechanics between Bohr and Einstein, initiated at the 1927 Solvay conference and pursued over several years, had not persuaded Einstein to join the herd of physicists who were convinced that the probabilistic interpretation of quantum mechanics was here to stay.

Einstein went on in his letter to make an important statement about singularities in field theory, and emphatically rejected the existence of singularities. At these mathematical singularities in his gravitation theory, the density of matter becomes infinite and his field equations are no longer valid. By implication, Einstein was rejecting the prediction by his own gravity theory of black holes, which contain singularities lurking at their centres. Moreover, he was also rejecting the "Big Bang" model of cosmology also based on his gravity theory, in which a singularity at time t equal to zero must inevitably occur. What Einstein actually wrote to me about this was:

I only want to point out that Infeld's objections are not justified. This is because they assume that even for non-gravitational interactions between systems, the areas of weak fields are dominant. The quantum facts teach [us] that in truth this cannot be. A complete field theory cannot allow any singularities.

Einstein's reference to Leopold Infeld, his former collaborator at the institute at Princeton, concerns a paper published by Infeld criticizing Einstein's unified field theory. The paper claimed that the theory cannot produce the correct motion of a charged particle in an electromagnetic field, otherwise known as the Lorentz equations of motion. In one of the papers that I had sent to Einstein, I had discussed this problem of the motion of charged particles in his unified field theory.

Einstein closed his letter with some insightful observations about quantum mechanics, which are very relevant to present-day attempts to understand the foundations of quantum mechanics and to quantize Einstein's gravitation theory.

Naturally it is quite possible that it is not possible at all to do justice to reality with a field theory. Then, however, in my opinion, one is not allowed at all to introduce the continuum (also not the "space") and I see in this circumstance no concepts on which one can rely with some prospect of success.* At any rate, I do not put any hope in subsequent "quantization." But all this does not claim to be objectively correct. I simply see it this way.

Einstein signed his letter "friendly greetings to you, Your A. Einstein." Although he had not directly addressed my question about my abilities as an aspiring physicist, he wrote to me as an

---

*Einstein's gravitation theory is based on the notion that spacetime is continuous. Many attempts to quantize the gravitational field have been based on such a "continuum." However, in recent years attempts have been made to make spacetime and gravity emerge from a discrete, quantum description of spacetime geometry.

intellectual equal, which astonished me and certainly strength-ened my motivation to become a physicist.

When the barber finished translating the letter as well as he could, he handed it back to me and said softly, "Well, John, it looks as if Herr Doktor Einstein is taking you seriously."

*

In my next letter to Einstein, on July 21, 1953, I attempted to address some of his profound statements about how physics should be pur-sued. I wrote, in effect, an essay on the epistemology of physics and whether one should do physics from a top-down or a bottom-up approach. The top-down approach attempts to do physics based on a priori reasoning, with the aim of achieving an elegant, beautiful theory with the least number of fundamental assumptions; the bottom-up approach is based on experimental facts and builds up from these facts to a consistent theory. For example, modern par-ticle physics usually takes a bottom-up approach, in that particle theories are developed on the basis of known experimental data. Current top-down theories are, for example, string theory and quan-tum gravity. In my letter to Einstein, I also mentioned the need for creative ideas in physics and described how a theory develops from imagination, while in the end it is necessary to confront it with experimental observations.

In his letter to me, Einstein had been concerned about whether quantum mechanics was indeed a complete description of reality; he had contended it could not provide a useful starting point for a more profound theory. Einstein held this view in spite of the empir-ical successes of quantum mechanics: the theory agrees with all the data from subatomic systems, and there is no known experiment that contradicts it.

In my answering letter, I discussed the pivotal issue of determin-ism versus non-determinism, where in classical physics the position and speed of a particle can both be determined with infinite accuracy, but through Heisenberg's uncertainty principle, this cannot happen in quantum mechanics. Quantum mechanics is a non-deterministic system of physics, for it is based on statistical probability theory. I wrote:

> I feel subjected to the same indecision as Buridan's ass, which was unable to choose any specific bundle of hay.* This so with respect to the controversy Determinism *versus* Indeterminism. In spite of this, my *intuition* tells me which specific bundle of hay it is most propitious to decide on. The scientific minded youth of today see this dilemma in a different light to those who developed and lived with the problem. Consequently, they do not realize any problem whatsoever; they believe the aim of Scientific Comprehension is "the second principle" (Indeterminism) in the limiting realm of the "quanta." This conception (purely conventional) is typical of the age; owing to this, I do not believe in it as final.

What I was aiming to say to Einstein was that the younger physicists of the time simply accepted quantum mechanics without questioning whether it was a final and complete description of reality. However, today, more than fifty years later, the tide has turned, and many physicists are thinking about the foundations of quantum mechanics and questioning whether the present so-called Copenhagen interpretation of quantum mechanics is viable.

---

*The parable of Buridan's donkey, or ass, satirizes the moral determinism of the fourteenth-century French philosopher Jean Buridan. Placed between two bales of hay, the donkey dies of hunger because it can't decide which to eat. Another version has the donkey dying of both hunger and thirst, unable to choose between a bundle of hay and a pail of water.

In my letter to Einstein, I also discussed the importance of empirical verification of a physics theory, quoting from Lord Rutherford: "It seems to me *unscientific* and also dangerous to draw far-flung deductions from a theoretical conception which is incapable of *experimental* verification, either directly or indirectly." As a twenty-year-old, I anticipated a basic problem that faces physics today, namely the difficulty in obtaining sufficient experimental data to verify and test such theories as string theory, quantum gravity and other highly speculative theories in cosmology. The only way to obtain new data is through increasingly large and expensive high-energy accelerators, which is leading to a crisis in physics today.

I then wrote about theory versus experiment. ". . . It is always possible to modify a theoretical scheme (by additional artificial assumptions) in such a manner as to obtain *immediate* experimental verifications . . . One shall not modify the true aim of science (aim of complete comprehension) for the sake of momentary interests. However, if this is found necessary, the step shall only be understood as a temporary state of affairs . . ." Today, this problem of ad hoc physics is even worse than in Einstein's day. Today we have the speculation that exotic "dark matter" and "dark energy" exist in order to explain astronomical and cosmological observations that do not follow from Einstein's gravity theory.* Physicists today have "modified" Einstein's gravity theory by adding in the "artificial assumption" of undetected exotic dark matter to "obtain immediate experimental verification."

---

*Most physicists today believe that 96 percent of the matter and energy in the universe is invisible. According to the standard model of cosmology, about a third of this invisible energy and matter is called "dark matter," and the rest is in the form of "dark energy." Without dark matter, the standard model of Einstein's general relativity would not explain the data showing that there is stronger gravity (i.e., more matter) in galaxies and clusters of galaxies than is observed. Similarly, invoking an invisible and repulsive energy explains why the expansion of the universe appears to be accelerating.

Later in my eight-page letter, I confronted the problem of singularities in gravitation theory, which Einstein had discussed in his letter. I speculated on whether there might be a possible criterion for when the solutions in field theory and gravity theory are regular or non-regular (that is, non-singular or singular, respectively). I also discussed the issue of the cosmological constant as "corresponding to a universal field density."* I was proposing that this energy field associated with the cosmological constant, which today is interpreted as the universal vacuum energy, or "dark energy," had to be included in a truly unified theory. Yet Einstein did not like his cosmological constant because it introduced what he called a "heterogeneous piece" into his basic gravity equations.

Finally, I wrote about how gravity theories should be purely geometrical in origin, avoiding a phenomenological description of matter such as Einstein had used in his gravity theory and in his first paper on cosmology, "Cosmological Considerations in the General Theory of Relativity," published in 1917.** Einstein addressed the issue of a purely geometrical theory of gravity in a well-known quote: "Gravitational equations of empty space are the only rational well-founded case of a field theory." He meant by this that the right-hand side of his field equations for gravity should be zero,

---

*Einstein introduced the cosmological constant, λ, in 1917 to prevent the natural expansion and contraction of the universe that followed from his field equations. It provided a repulsive gravitational force that made the universe static. Astronomers in his day accepted that the stars and nebulae were static, and for once, Einstein was going along with the herd. Later in life, after the astronomer Edwin Hubble discovered that the universe was expanding, Einstein told the Russian-American physicist George Gamow that introducing the cosmological constant was the biggest blunder of his life.

**The phenomenological part of Einstein's theory is described on the right-hand side of his field equations. "Phenomenological" is a derogatory term sometimes used by physicists to describe an ad hoc explanation for a physical phenomenon that is not founded in fundamental physics.

not the phenomenological energy momentum tensor postulated in his papers on his general theory of relativity. Already early on in his research on gravity, Einstein was dissatisfied with the formulation of general relativity. Einstein was never entirely happy with the research he published. He was always looking ahead, ambitiously, to a more fundamental unified description of physics.

I didn't wait to hear back from Einstein, but I wrote to him again, on August 12, and this time my letter was much more technical. I also enclosed a manuscript I had just completed, entitled "Unified Field Theory." In the letter and in my paper, I discussed the fact that in Einstein's attempts to construct a unified field theory, he had not included a field associated with the strong and weak nuclear forces. These forces were already known to be important in the early 1950s, and had been discovered by observing particle collisions in early accelerators. I really felt that Einstein needed to include these forces in his theory. It was clear to me that without them, he could never hope to achieve a correct and complete unified field theory.

Einstein must have responded almost immediately to this letter, for his next one to me was dated August 24. "Dear Mr. Moffat!" he began, in another handwritten letter:

Our situation is the following. We are standing in front of a closed box which we cannot open, and we try hard to discuss what is inside and what is not. The similarity of the theory [his unified field theory] with the one by Maxwell [electromagnetism] is only superficial. Thus we cannot simply take over the concept of a "force" from this theory to the asymmetric field theory. If this theory is useful at all, then one cannot separate the particle from the field of interaction. Also there is no concept at all of the *motion* of something that is more or less rigid. The question here is exclusively: Are there solutions without singularities? Is there energy preferably localized in such a way as it is required by our knowledge of the atomic and quantum character of reality? The answer to these questions is indeed not achievable with present mathematical means. Thus I do not see how one should suspect whether some remote action and some objects, as far as we have gained a semi-empirical knowledge of them, are represented by the theory. Thus our pertinent complete ignorance does not begin with "nuclear forces." Here the situation is different from the pure theory of gravitation, where one can approximate the masses through singularities.

Einstein was willing to accept that he was ignoring the nuclear forces. However, he was trying to justify this omission by saying that we are not able to understand these nuclear forces with the mathematical tools available at that time.

"The only thing that is in favor of the new theory," he continued, still referring to his unified field theory, "is the fact that it appears to be the only natural generalization of the equations of the pure gravitational field."

I wrote two more letters to Einstein, enclosing further calculations based on his nonsymmetric unified field theory. I discussed some critical aspects of his field equations. He responded in October 1953 with a short note, advising me to be careful about publishing this work prematurely. I took his advice and continued developing his theory.

*

When I wrote to Einstein in 1953, he was one of the most celebrated physicists in the world, and had won the Nobel Prize long before, in 1921.* But he had isolated himself from the rest of the physics

---

*The Royal Swedish Academy of Sciences actually did not award a physics prize in 1921, but awarded Einstein the 1921 physics prize in December of 1922.

community by his problematic stand on quantum mechanics. He felt that "God does not play dice with the universe," a metaphorical swipe at the random, probabilistic nature of quantum mechanics. His disagreements with Niels Bohr on the interpretation of quantum mechanics were legendary. His work on unified field theory and gravitation was disconnected from the mainstream of physics, which at that time was concentrating on developing nuclear physics and atomic physics. Most other physicists dismissed Einstein's attempts to find a generalization of his gravity theory that unified it in a geometrical framework with James Clerk Maxwell's equations for electromagnetism. Indeed, Einstein was ostracized by his physicist peers.

This ostracism began as early as the 1930s, when Einstein appeared before a committee at the Institute for Advanced Study, requesting financial assistance to bring Leopold Infeld from Poland to the institute to assist Einstein in his calculations of the motion of particles in his gravitation theory. He was denied this request. This prompted Einstein and Infeld to write a bestselling popular book together, titled *The Evolution of Physics*. Money from the sales of this book paid for Infeld to travel to Princeton and begin the collaboration. Much later, in the late 1940s and early 1950s, when Robert Oppenheimer was director of the institute, he dissuaded the young physicists there from associating with Einstein because he believed that Einstein would be wasting their time; he didn't want Einstein influencing the younger generation.

Is it possible that Albert Einstein showed interest in corresponding with me, an unknown, self-taught student of physics in Denmark, because I had shown critical interest in his work? Unlike his esteemed peers, I, a young aspiring physicist, was taking him and his recent work seriously. Also, in opposing the consensus view of what physics was worth doing and what was not, I had positioned myself as an outsider at the Niels Bohr Institute circle, just as Einstein himself had become an outsider.

Ultimately, Einstein's attempts to unify gravity and electromagnetism were not successful. However, he was the first to modify his own theory of gravity, and, in fact, he was many decades ahead of his time in not accepting his gravity theory as the final truth. Today there are many published modifications of Einstein gravity, because physicists want to understand the nature of dark energy. Modifying Einstein's gravity theory may be able to avoid both this undetected dark energy and the need for Einstein's problematic cosmological constant. Similarly, the concept of "dark matter"—and the problem of why most of the matter in the universe is "invisible"—has led to other modifications of Einstein's gravity theory, as some physicists attempt to explain the dynamics of galaxies and the large-scale structure of the universe without having to postulate the existence of undetected, exotic dark matter. The work that I began as a twenty-year-old when corresponding with Einstein has evolved into my modified gravity (MOG) theory, which I am still developing today.

As for the debate between Einstein and Bohr over the interpretation of quantum mechanics, most physicists today contend that Bohr was right in believing that quantum mechanics was a complete description of reality. However, there are some leading physicists who claim that quantum mechanics may not be a complete description of reality, but rather an underlying statistical description of nature, which will eventually be encompassed by a more complete, deterministic theory.

5

―――

ERWIN SCHRÖDINGER

N OT LONG after my discouraging talk at the Niels Bohr
Institute, the correspondence with the British consulate
bore fruit. In the meantime, the consular officials had
learned that I had indeed been interviewed by Niels Bohr, had
given a talk at the Niels Bohr Institute and had been corresponding
with Einstein. I received a letter from a Mr. Greenall, who was a
scientific attaché in the Department of Scientific and Industrial
Research in London. He invited me to visit the department in London, suggesting that they could assist me in contacting physicists at
universities in England, which might lead to my being able to continue my physics research in an academic environment. I was filled
with anticipation and enthusiasm upon receiving this letter, but
at the same time I was not surprised. I felt as if things were evolving according to my hopes and expectations.

The government department paid my fare from Copenhagen to
London, and I sailed across the North Sea from Esbjerg port in
Denmark to Harwich. It was January 1954, the voyage was in rough
weather, and the Danish meal I'd had before departure, courtesy of
the British government, did not sit well at all. Upon my arrival in
London, already displaying a flair for high living, I checked into the

Regent Palace Hotel in Piccadilly Circus, putting the bill for the hotel room on the expense account that Greenall had provided for me.

The next day I visited the offices of the Department of Scientific and Industrial Research, located on Oxford Street not far from my hotel, and was greeted by a cheerful secretary who offered me a cup of tea and biscuits, explaining that Mr. Greenall would be in his office in half an hour. We chatted about my rough trip across the North Sea. She asked me where I was staying and when I said the Regent Palace Hotel, she laughed, which puzzled me.

Mr. Greenall, a tall man in a dark suit, stiff, white-collared shirt and what looked like an old school tie, finally arrived. He took off his bowler hat, revealing his large bald head. He ushered me into his office, invited me to sit down and asked the secretary to also bring him a cup of tea and biscuits. I now recollected that much of what went on in English circles was accompanied by tea and biscuits. Mr. Greenall was friendly as we chatted about the rough North Sea voyage, but when he learned that I was staying at the Regent Palace Hotel in Piccadilly Circus, his friendly smile faded.

"Oh, dear," he said, "we can't have this. Far too expensive for a young man like you. We'll have to find you alternative accommodation in London." He looked at me penetratingly, and I wondered whether he was regretting having brought this brash young genius to London after all.

"So I understand you have corresponded with Albert Einstein," Mr. Greenall began, after the hotel arrangements had been cleared up.

"Yes," I said. "I've written two papers on his unified theory, and that was what the correspondence was mainly about."

"So you want to pursue a career as a physicist?"

"Yes, I would like to get into an academic environment."

"I hear from the British consulate in Copenhagen that you were interviewed by Professor Niels Bohr, and he seems to have taken an interest in your future. He suggested to the consulate that you should pursue an undergraduate education, so that you could be trained properly as a physicist. What do you think about that?"

I recrossed my knees and looked out the window. It had started raining. The secretary came into the office, removed the teacups and smiled at me, and I smiled back. She seemed to have taken a motherly interest in me.

"Well," I began, feeling awkward about contradicting the great Niels Bohr, but greatly desiring to move forward quickly, "I've already learned the equivalent of an undergraduate degree in physics and mathematics myself in Copenhagen, and I want to continue my research on Einstein's unified field theory and other subjects in theoretical physics."

Greenall looked at me with surprise and said, "Do you really think that you can pursue an academic career without an undergraduate degree?"

"Yes, I think it would be a waste of my time to spend four years relearning what I already know." I was feeling worried, too, about perhaps being forced into a conventional route in academia, which could, through the typical rote learning, extinguish my creative impulses and my strong motivation to pursue my own path.

Greenall seemed dubious and stared at me intently for a while. "What I plan to do," he finally said, "is to arrange interviews for you with various well-known physics professors here and see what they think about how you can best continue your research and achieve your goals. We will pay your expenses visiting these professors. I have a list of them. The first one you should visit is William McCrea at Royal Holloway College, not far from London. He's an expert on relativity theory. After that, we'll send you to Liverpool. Professor William Bonnor has expressed an interest in talking to you. He's a mathematical-physics professor at the university and has published on Einstein's unified theory. Then we'll send you to Dublin,

Ireland, to the Institute for Advanced Studies, where I've arranged for you to meet with Professor Erwin Schrödinger, who's the director of the institute."

"Aha!" I exclaimed. "I've read papers by Professor Schrödinger on unified theory, and he has his own way of attempting to unify gravity and electromagnetism. It would be exciting to talk to him!" Schrödinger was one of the founders of quantum mechanics and had won the Nobel Prize in 1933, jointly with Paul Dirac, for developing this theory. I was thrilled that this would be the second founder of quantum mechanics I would have the opportunity to meet and talk to.

Greenall seemed satisfied with my response to the Schrödinger interview and said, "Crossing the Irish Sea in the wintertime can be a stormy affair. I hope you have an easier time than you did crossing the North Sea. You can stay another night at the Regent Palace Hotel, and I will phone tomorrow with suggestions for cheaper rooms in London. Once you've found appropriate accommodation, you can begin your visits to the universities."

I left Greenall and the cheerful secretary, and the next day I began hunting for accommodations in London using Mr. Greenall's list, regretfully leaving the luxurious Regent Palace Hotel, where I had enjoyed a final evening in the grand restaurant listening to a musical ensemble. I located a room in a cheap residential area on the outskirts of London, which I could reach by the London underground, the tube. This would be my base in London for a couple of weeks while the secretary at the government department arranged for train and boat tickets. I would have breakfast with the landlady and her husband, and would have to put up with the strong smell of cat urine that permeated the house, thanks to their tomcat, Billy. Greenall sent a message to me at my new address saying that he had now arranged a further visit to Cambridge University after my visit with Schrödinger in Dublin; I was to meet with a Dr. Dennis Sciama at Trinity College, Cambridge.

*

Professor William McCrea had published a book on special relativity, which I had read during my library studies in Copenhagen. He had also published papers on cosmology. He was head of the Physics Department at Royal Holloway College, then an all-girls college. My visit with him was cordial, although he seemed rather skeptical about my proposed plans for doing research at an English university without an undergraduate degree. He said that he was not an expert on Einstein's unified theory, whereas Bill Bonnor at the University of Liverpool had published several well-known papers on the subject and he would be a better person to advise me on my future.

A week later, I travelled to Liverpool by train and visited Bonnor, who was in the Department of Applied Mathematics. We talked about some of the technical aspects of my papers on Einstein's unified field theory and Einstein's reactions to them. Bonnor knew a lot about Einstein's work on unified theory, and described papers that he had written solving Einstein's field equations, which had been published in the *Proceedings of the Royal Society*. I had actually read his papers while I was studying Einstein's unified theory, so I was able to make reasonably intelligent comments about his work.

The next day, I departed from Liverpool on an overnight steamer to Dublin. As Greenall had predicted, we ran into a storm on the Irish Sea and the crossing was very rough. However, the stormy trip was made more pleasant by my meeting an Irish nurse who worked in a hospital in London and was on her way home to Dublin to visit her family. She was just as seasick as I was.

Greenall had arranged a room in a quaint hotel near the river Liffey, and from my window I could see the famous Guinness brewery across the river. The next morning, I went to the Dublin Institute for Advanced Studies, which was a small greystone building not far from the centre of Dublin. A secretary said that I should wait for Professor Schrödinger's wife, who would drive me to their house outside Dublin. In the meantime, she said, I could talk to Professor Synge, who had published a book on Einstein's relativity theory, and was related to the famous Irish playwright John Millington Synge.

She took me to an office where a smiling, middle-aged man was standing at a blackboard discussing physics with a younger man who turned out to be Bruno Bertotti, an Italian researcher at the institute. After we had shaken hands, Synge said, "We're trying to do this integral. Maybe you can help us." He handed the chalk to me, and I suddenly felt dismayed that I would have to perform for these two physicists.

I looked at the blackboard, rubbed my chin and said, "I think I can do this integral," and proceeded to work it out.

As I finished, Synge said, "Well done, well done, young man! I hear that you are here to visit Erwin and talk about his work on unified field theory."

I said, "Yes, I'm looking forward to speaking to him about my work and his work."

"I dare say that should be interesting," he said, his eyes twinkling.

Bertotti, a studious-looking young man wearing Schubert spectacles, offered, "Physicists today express the opinion that Einstein is wasting his time with this unified field theory business."

Just then the office door opened and in walked a tall woman in a black leather coat and aviation helmet. Synge introduced me to Mrs. Anny Schrödinger. As we shook hands, she said, "Let's leave immediately so that we can get home in time for lunch."

We drove away in an old black Mercedes-Benz, going around the bay outside Dublin to the Schrödinger home, which was a modest and unimposing townhouse, joined on either side to others in an unextraordinary suburb of Dublin.

*

Upon entering the house, Anny Schrödinger introduced me to another, younger woman, who came out of the kitchen. I later learned that she was actually another liaison of Schrödinger's, for he was known to have mistresses with the knowledge of his wife. I also learned much later that Anny, for her part, had had a long-term affair with Hermann Weyl, who was one of the foremost mathematical physicists of the twentieth century.

Anny said, "Professor Schrödinger is not well and has to stay in bed. We will have lunch, and you can speak to him afterwards."

The Schrödinger house was cramped and cluttered: stacks of books and piles of newspapers occupied most surfaces and corners. After lunch, which the three of us ate together in the dining room, the other woman busied herself in the kitchen while Anny and I talked about my situation and the reason that I had come to meet Professor Schrödinger. She appeared to me to be a very intelligent woman with firm views about life.

After our chat, she took me upstairs to the top of the house. We entered a small bedroom and Anny introduced me to Erwin Schrödinger. The legendary physicist made an incongruous figure, sitting up in bed in a small alcove, wearing a large white nightshirt and a crocheted woollen bed hat. As we shook hands, I noticed that his hand felt clammy and frail.

"So you are Moffat, the young man who has been writing papers about unified field theory and corresponding with Albert. Do you have copies of the papers that you sent to Albert?" Schrödinger asked, speaking English with a light German accent.

"Yes," I said. I pulled out the papers from my bag and handed them to him, just as I had done with Niels Bohr a few months previously. Still sitting up in bed with his woolly hat on, Schrödinger started reading the papers, his reading glasses astride his beaky nose. All of a sudden he was overcome by a coughing fit. He looked at me with reddened eyes and said, "I've had this bronchitis. You understand this is why I am in bed." I made a sympathetic response. He returned to my papers, and I sat looking around his little room.

It was even more disorganized than the downstairs rooms. Stacks of books reached up to the ceiling, and untidy piles of papers covered a small writing table. It looked to me as if this was his room only; the bed was a single bed, and there were no feminine articles to be seen. It was amazing to me to ponder that I, two years ago merely a messenger boy and window cleaner, was now sitting in the bedroom of one of the greatest physicists of all time, a Nobel Prize winner, and one of the founders of quantum mechanics.

Schrödinger had invented wave mechanics and what is now called the Schrödinger wave equation, which was compatible with Werner Heisenberg's matrix mechanics.* Legend had it that he wrote the first important papers developing his quantum wave mechanics while vacationing in the Alps with one of his mistresses, but no one has been able to discover the name of this muse. In addition to all his famous papers on statistical mechanics and colour theory, he also wrote a wonderful book on relativity called *The Structure of Spacetime*, which was one of the treasured books from which I had learned the basics of relativity and gravitation just over a year before. In 1944, he had even published a groundbreaking book on molecular biology that anticipated the discovery

---

*The wave equation named after Schrödinger describes the wave nature of the electron as opposed to the particle nature of the electron in Heisenberg's matrix mechanics. In quantum mechanics, the electron has the dual nature of a wave and a particle, and both Schrödinger's and Heisenberg's interpretations are equivalent.

of DNA. I learned that he spoke six languages, including English and German. He had left Austria when he was a professor at the University of Graz because he was concerned about the spread of Nazism, and he had ended up in Dublin. At the time that I met him, Schrödinger was in his late sixties and was about to retire as director of the School of Theoretical Physics at the Dublin Institute for Advanced Studies, which he had helped to found.

Schrödinger finished reading the papers, looked up at me sternly and said, "I see that you are using Albert's method of deriving the unified field equations from a variational principle." He was referring to an important mathematical derivation of equations that would describe the unified field theory of gravity and electromagnetism. "Why are you not using my method of deriving these equations?" he continued irritably.

I suddenly felt nervous, and saw my precarious future in physics threatened. "Are you referring to the method you describe in the appendix of your book on spacetime structure?" I asked.

Schrödinger leaned forward and looked at me even more sternly. "Yes, yes, that is what I am referring to. This is the obvious way to derive the unified field equations."

"But don't they lead to the same answer?" I asked innocently.

Schrödinger straightened up in the bed and his face flushed. "Well, yes, they may lead to the same result, but my method is far superior to Albert's! Let me explain to you, Moffat, that Albert is an old fool."

I said, "I beg your pardon?"

"He's an old fool," Schrödinger repeated, "and you should not be following his methods. You should follow what I do because I know how to derive these equations properly."

I sat stunned in my chair. During the interview with Niels Bohr, Bohr had accused Einstein of being an alchemist and wasting his time. Now another equally famous physicist was accusing Einstein, who had been so kind to me, of being an old fool. What was I supposed to make of this?

A few awkward moments of silence ensued. Suddenly there was a loud crash of porcelain breaking downstairs in the kitchen. Then a spate of angry female voices floated up the stairs. Schrödinger looked irritably at the open door to his room, and then looked back to me and said, "Well, what do you have to say about this?"

I knew that this interview was very important and would decide whether I was going to have a chance for a future in physics or whether I would find myself back in Copenhagen performing least squares calculations at the geophysics institute. I felt like a drowning man whose life is passing by before him. I thought, "I have to say something to assuage his feelings about this." Later in life, I learned that I had been caught in a crossfire between Schrödinger and Einstein. A war of words had erupted after Schrödinger published a paper about his unified theory in the *Proceedings of the Royal Irish Academy*, a journal that published, among other things, Gaelic poetry. Schrödinger had let it be known to the international newspapers that he had discovered the true unified theory of gravity and electromagnetism. Learning about this in Princeton, Einstein accused Schrödinger of plagiarism and threatened to sue him. Schrödinger counterclaimed that Einstein had plagiarized *his* work, and threatened a countersuit. Wolfgang Pauli, another founder of quantum mechanics, had to intervene, and managed to cool down the tempers on both sides. I realized that physics was not the pristine endeavour to find the truth that I had idealized as a young man, but it was cutthroat and competitive, and even those who were world-famous were still grasping for even more immortality.

"Well, Professor Schrödinger," I replied carefully, "if I proceed further with my work on unified theory, I certainly plan to use your method of deriving the unified field equations."

Schrödinger suddenly smiled and said, "Well, that's fine, young man. We'll let it pass."

We talked for a while longer about the future of unified theory, and the possibilities of verifying or falsifying such a theory experimentally, a prospect that did not seem very promising at the time, and in fact hasn't improved today. We ended the discussion cordially and I shook his hand. He had another coughing fit and said he needed to rest. I went downstairs and Anny offered tea, which I accepted. Before driving me back to my quaint hotel in Dublin, she took me out into the garden at the back of the house for a tour of her flower beds.

*

As it turned out, Schrödinger wrote a positive letter on my behalf to Greenall in London, and as far as I know, so did Professors Bonnor and McCrea. I believe that Schrödinger's letter would have been the important catalyst that led to Greenall continuing to provide me with financial support and promoting my visit to Cambridge University to meet Dennis Sciama.

About a week later, I took the train up to Cambridge from London and met with Sciama at Trinity College. Walking through the Great Gate at Trinity and entering the Great Court, I was impressed with the university surroundings and Trinity College itself. I felt in awe to be standing in the same college where Isaac Newton developed his gravity theory and wrote his great treatise, the *Principia*, and where many other famous scientists and scholars had produced great works over the centuries.

Dennis Sciama was twenty-eight years old when I met him in his rooms at Trinity, where he was a fellow. He was doing research on Mach's principle, attempting to incorporate it within a theory of cosmology. The nineteenth-century philosopher-physicist Ernst Mach had the idea that the inertia of a body, which is its resistance to an external force, owed its existence to all the rest of the matter in the universe. A problem with this idea is that it is clearly difficult or impossible to ever verify, for we cannot remove all the rest of the matter in the universe to discover whether the inertia of a body still exists. Sciama was attempting to explain the inertial force of a body by means of a variable gravitational constant, G, following his supervisor, Paul Dirac, who proposed that Newton's gravitational constant varies in space and time. Dirac and Sciama's work was the forerunner of later work by Pascual Jordan, Carl Brans and Robert Dicke connecting a variable G with Mach's principle and the origin of inertia.

Later in his life, when Sciama was a lecturer at Cambridge, he supervised several students who became leading relativists and cosmologists, including Stephen Hawking, George Ellis, John Barrow, Gary Gibbons and Brandon Carter, as well as the astrophysicist Martin Rees.

When I visited him, Sciama had rooms under the big clock in the Great Court. He was an intense, handsome young man. His spacious rooms suggested a wealthy family background. We talked about my correspondence with Einstein, my meetings with Niels Bohr and Erwin Schrödinger, and my desire to become a physicist. Sciama must have been impressed by the letter that Schrödinger wrote, because without further ado, he said, "Come with me. I'll see that you are matriculated, that is, enrolled, at the college this afternoon. I understand that you don't have an undergraduate degree."

I said no, and explained how I had learned physics and mathematics on my own.

"I am matriculating you without requiring you to take the tripos exams because I think that you will be capable of doing a Ph.D.," he said. I was relieved to know that I could go straight into a Ph.D. program without having to endure four years of undergraduate training leading to these infamous, gruelling exams. "I've been in contact with Fred Hoyle," Sciama continued, "who has agreed to supervise you towards your Ph.D."

He took me to an office at the college and he spoke to a clerk, who made up the papers for me to be enrolled as a student at Trinity. The year was 1954.

How much was my own hard work and how much was luck? In arriving at this turning point in my life, many kind strangers—famous physicists as well as influential bureaucrats—had encouraged me and opened doors for me. I signed the papers at Trinity, shook Dennis Sciama's hand and started out in earnest on the path that I had chosen, or perhaps the path that had chosen me.

6

FRED HOYLE

WITHIN A WEEK I had moved up to Cambridge and found a room in a house in the Fens, so called because it was originally a marshy area outside Cambridge. It was still winter, and cold and damp. In my room was a monstrous gas fire, controlled by a large metal box, which gobbled up all the shillings I could feed into it. I had to make regular trips to Barclay's Bank on Benet Street, where I had opened an account, in order to acquire more bags of shillings. Being used to modern heating appliances in Denmark, I soon moved my studies to the library above the Arts School on Benet Street, where the Theoretical Physics Department was located, as well as the Mathematics Department. There I would sit every day in relative warmth, clad in my black gown, which was mandatory apparel for students when in official university buildings and on the grounds. I was surrounded by other black-clad students seeking warm working conditions. The walls of the library were lined with hundreds of books in mathematics and physics, and the chief librarian in his grey smock was a friendly man who was helpful in retrieving the papers that I needed for my research.

At lunchtime, I would walk down King's Parade to the Great Hall at Trinity College, still wearing my black gown. I felt quite lonely since I had not yet managed to get to know any fellow students. In the library, everyone kept to themselves, studying intently and in silence. The quality of the lunch at college left a lot to be desired. It usually began with soup. One day I accidentally splashed my Trinity College tie with the soup, and soon a hole appeared there. What was the soup doing to my stomach! The dining hall at Trinity was one of the largest and most impressive of any of the colleges at Cambridge. There was a lounge for students on the second floor with a portrait of Sir Isaac Newton peering down at us from the wall, along with other notable Trinity scholars. I also visited the Wren Library, which had valuable physics manuscripts and had acquired a musty smell from the thousands of ancient books in the library stacks.

A few days after arriving at Cambridge, I saw Fred Hoyle in his rooms at St. John's College. Hoyle, who was about forty at that time, was quite congenial and asked me to sit down in a chair facing his desk with a large oil painting of himself behind it on the wall. I wondered whether all his students were forced to sit during tutorial sessions and contemplate Hoyle's blunt Yorkshire features in duplicate while he sat in an easy chair nearby.

Everyone at Cambridge referred to him as "Mr. Hoyle" because he had never felt the need to obtain a Ph.D. It was not uncommon for well-known academics in science in those days to avoid obtaining a doctorate. Hoyle was the only person at Cambridge besides Dennis Sciama, and a visiting physicist named Felix Pirani, who was actively engaged in research on relativity theory and cosmology, and since I was now already known around the university as the person who had corresponded with Einstein and worked on unified field theory, Hoyle was an obvious choice for my supervisor. At the time, Hoyle was working in collaboration with Margaret and Geoffrey Burbidge and William Fowler at Caltech on the origin of matter in stars and the early universe. The collaboration produced a famous paper in 1957, which derived the abundances of nuclear elements in the early universe and described the production of heavy elements in stars. The paper, entitled "Synthesis of the Elements in Stars," became so famous that it was widely referred to as "B²FH," meaning "Burbidge squared, Fowler, Hoyle."

Relativity and gravity were not popular research topics in the 1950s. Most of the activity in fundamental theoretical physics was concerned with atomic and nuclear physics, and also the very active field of particle physics. There was a dearth of experimental evidence for the validity of Einstein's gravity theory. There were only three classical tests of the theory: the bending of light, which had been verified in 1919 and, more convincingly, at the solar eclipse of 1922; the perihelion advance of Mercury;* and the difficult-to-verify observations of gravitational red shift by the sun and the dwarf star Sirius B. In contrast, the more popular field of particle physics had a wealth of experimental possibilities at the high-energy accelerators that were being built around the world. Indeed, the editor of the prestigious American physics journal *Physical Review*, Samuel Goudsmit, had threatened to ban all papers on gravitation in the journal. Since I had been actively pursuing unified field theory, which involved gravitational research, I wanted to continue this research at least for the first year or two at Cambridge, even though it wasn't in the current mainstream of physics.

---

*The nineteenth-century French astronomer Le Verrier discovered that the planet Mercury's orbit did not agree with the predictions of Newtonian gravity. The perihelion, or position of the planet's orbit closest to the sun, advanced over time to form a rosette pattern. Le Verrier concluded that a new planet must be responsible for this gravitational anomaly, and must lie between Mercury and the sun. He christened the undetected planet "Vulcan." But in 1915, Einstein calculated the perihelion advance of Mercury using his equations of general relativity and found it agreed accurately with over a century's observational data, without any need for a new planet. The reason for the anomaly in Mercury's orbit turned out to be the warping of spacetime near the sun, Einstein's new concept of gravity.

At our first meeting, Hoyle wanted to discuss my future right away, and I related to him the usual story of the Einstein correspondence and how I was working on Einstein's unified field theory.

As I anticipated by now, Hoyle said, "Well, John, in my opinion, Einstein is wasting his time with this unified field theory business. For that matter, so is Schrödinger." This was delivered matter-of-factly in Hoyle's thick Yorkshire accent.

I thought it prudent to remain silent.

"It's my understanding that you don't have undergraduate training," he said, changing tack.

"I've discussed this with Dennis Sciama," I replied. "As you know, he matriculated me without an undergraduate degree."

Hoyle eyed me through his thick-lensed glasses, and after a few moments of hesitation, he said, "Perhaps you should consider taking the tripos exams, and perhaps even spending four years studying for the Mathematical Tripos Part II, before continuing your physics research. After all, a physicist needs a proper training to be able to produce useful research that can make a mark on the physics community. Don't you think you should do this?" I shuddered at the thought.

"Personally I think I'd be wasting my time. I would prefer to continue working towards my Ph.D. because I believe I have enough technical ability to succeed."

Hoyle leaned back abruptly in his chair and didn't look convinced. "You realize you're taking a chance, and you could well fail?" he continued. "I think you should contact Dr. Felix Pirani who is presently visiting Cambridge and doing research in relativity theory. He can advise you on your best course. He is a very competent physicist, well versed in relativity theory."

I wondered why Hoyle was passing some of the responsibility for me onto the shoulders of Felix Pirani, and I began to feel uneasy about my future at Cambridge.

*

I was officially associated with a tutor at Cambridge who was responsible for my well-being at the college. Four other first-year students and I met him in his comfortable rooms at Trinity one evening, where we stood around sipping sherry. The tutor informed us about various issues regarding college life and the rules that we had to abide by. In particular, we were told to wear our gowns in the evenings in order to distinguish ourselves from Cambridge residents, which emphasized the ritualistic class separation of "town" and "gown."

I had a problem that I suspected the other students in the tutor's rooms did not have: I needed to find financial support because the money from the British government would come to an end within a month. Mr. Greenall had advised me by letter to approach the Nuffield Foundation in London, which provided student scholarships through the Massey-Harris-Ferguson farm equipment foundation in faraway Canada. I arranged an interview by letter with a Mr. Sanderson, the assistant director of the Nuffield Foundation, and took the train to London to meet with him. He was sympathetic to my situation and, as others had been, was impressed by my correspondence with Einstein and the letter of recommendation from Schrödinger. Within two weeks, I was informed that I had been awarded a Massey-Harris-Ferguson Foundation grant amounting to ten pounds per week, which in those days was enough to keep a student alive.

I spent some time pondering my unusual and perhaps precarious situation at Cambridge, and began to form a plan to ensure that I would be left alone in the three or four years it would take me to finish my Ph.D. I decided to write more papers and submit them to the *Proceedings of the Cambridge Philosophical Society*, a prestigious old physics and mathematics journal. (I did not continue working on the papers I had sent to Einstein on his unified field theory. I had finally realized that his program for a unified field theory was a failure, and I wished to move on to my own attempts to modify Einstein's gravity theory.)

I spent many days at the Arts School library that winter, working on my first modified version of Einstein's gravity theory. I was trying to change Einstein's theory because I felt that there were deeper mathematical and physical aspects of his concept of gravity that had not been mined as yet. My idea was strikingly original. I had formulated a complex symmetric Riemannian geometry,* and the first paper was of a purely mathematical nature. I believed that this was possibly the simplest way to modify Einstein's theory of gravity and still remain within a purely geometrical structure of spacetime.

I knew that my supervisor, Fred Hoyle, was not terribly supportive of modifying Einstein gravity, either to develop a unified theory of gravity and electromagnetism or to simply modify relativity within the context of pure gravitation. In addition to his projects on the origin of matter in stars, Hoyle was busy working with his colleagues at Trinity College, Thomas ("Tommy") Gold and Hermann Bondi, on the steady-state universe cosmology, which was an alternative to the popular Big Bang model of the beginning of the universe. However, I was not interested in working on the steady-state model. The mathematical structure of my modified gravity theory was very appealing to me, and as it turned out many years later, it would have relevance for developments in quantum gravity theory.

To my delight, I received notice from the *Proceedings of the Cambridge Philosophical Society* that they had accepted my first paper on modified gravity. The journal editor suggested that I talk

---

*Riemannian geometry is a non-Euclidian geometry developed in the nineteenth century by Georg Bernhard Riemann that describes curved surfaces on which parallel lines can converge, diverge and intersect. Einstein made Riemannian geometry the mathematical formalism of general relativity.

with a mathematician at Trinity, who turned out to be an English baron. He was enthusiastic about my paper and only suggested one change: replacing a word that I had used incorrectly. I was pleased that it was only a mistake in English, not in the complicated mathematics I had used. Emboldened by this first success, I wrote two more papers, brazenly calling one of them "The Foundations of a Generalized Theory of Gravitation." Now I had a full-blown modification of Einstein's gravity theory. I submitted these papers to the same journal and they too were accepted for publication!

When he heard about the acceptance of these papers, Hoyle changed his mind about my needing to labour for several years in order to sit the horrendous tripos exams. In a subsequent meeting, I detected a shift in his attitude: from then on, Hoyle treated me more like a colleague than a student. For an hour, he would expound his latest ideas in his steady-state model, and describe his ongoing dispute with the Big Bang establishment. The steady-state model was based on the "perfect cosmological principle," originally proposed by Gold and Bondi. This principle states that not only does the universe look the same in every direction and matter is uniformly distributed throughout the universe, but it also looks the same at any *time* in the evolution of the universe. This is very different from the Big Bang theory, in which only the first two principles apply—known as the principles of isotropy and homogeneity.

It seemed to me that this perfect cosmological principle was indeed a simple, beautiful idea. In order to maintain it, Hoyle and his collaborators had published papers postulating that minuscule bits of matter were continuously created in voids in space. It didn't take the creation of much matter to continually refill the universe. In the steady-state model, there was no beginning to the universe, and yet it still had to explain Edwin Hubble's astronomical observations of the expansion of the universe and the recession of galaxies. Hoyle, Gold and Bondi postulated that, in order to conserve energy, matter was continuously created as the galaxies receded into the infinities of space. This was in direct contrast to the Big Bang model, championed by leading cosmologists such as Georges Lemaître and George Gamow, in which the universe had a singular beginning with an infinitely dense point of matter and, according to Lemaître, a huge explosion. Yet, ironically, it was Hoyle who gave what was originally known as the "dynamic evolving model" its catchy name. In the late 1940s, during a famous series of talks on the BBC, Hoyle dismissively referred to the primal explosion as the "Big Bang," and the name stuck.

During our tutorials, Hoyle wasn't really very interested in discussing the papers I was writing and publishing. He would sit in his easy chair under his portrait and talk rapidly about new developments in his steady-state model. He would stop his monologue occasionally and peer at me through his thick glasses and ask me a question such as, "Well, what do you think of these ideas about how to treat geodesic motion of particles in the steady-state model?"

My mind would race to find something intelligent to say, although I wasn't working on the subject. I would finally begin to sputter out an answer—"Well, I think that maybe you should . . ." Hoyle would then suddenly leap out of his chair, look nervously at his watch like the White Rabbit, grab his black gown and announce, "I have to leave! I mustn't be late for my eleven o'clock lecture!" We would then rush through St. John's College, out into Trinity Street and up the street to the Arts School where he was to give his lecture, without saying a word.

Following Hoyle's suggestion, I did seek out Felix Pirani, a scholar in his late twenties who was doing a second Ph.D. at Cambridge, having already completed one under the supervision of the Toronto relativist Alfred Schild. Felix lived with his wife in a house on the outskirts of Cambridge. He was also skeptical about my prospects of succeeding at Cambridge, and voiced the opinion that I should take the tripos exams. Indeed, during one of my visits to his house, he became quite voluble and aggressive on the subject. I became upset, the evening ended in a shouting match, and we parted on not-very-friendly terms. However, as with Hoyle, when I later informed him that my papers were being published in the *Philosophical Society Proceedings*, Pirani seemed to change his mind and we began to discuss serious issues in relativity theory together.

After my first year at Cambridge, with three published papers on my resumé, I was pretty much left to my own devices. I attended only two courses. I sat in on Paul Dirac's inspiring course on quantum mechanics and later decided to repeat the experience all over again. Dirac's exposition of quantum mechanics was like the construction of a beautiful symphony. It was so logical in its development that, despite Einstein's objections to the theory, Dirac's description convinced many physicists that quantum mechanics possessed its own beauty. The other course was on boundary value problems in potential theory given by an applied mathematician, Eggleston. This was an interesting subject to me in applied mathematics, which would stand me in good stead when I prepared my Ph.D. thesis on classical gravity theory.

*

One evening I attended a party at a big house at 9 Adams Road in an upscale residential area on the outskirts of Cambridge. This house enjoyed a certain notoriety in Cambridge circles, for its parties were an opportunity for young students to pick each other up, and engage in some serious drinking. The upstairs of the house was occupied by Professor Francis Roughton, a celebrated physiologist at Trinity. In the downstairs lived his ex-wife, Dr. Alice Roughton, an equally well-known family physician and psychiatrist, who regularly wore farmer's clothes and heavy boots that she had inherited from a former RAF pilot. She encouraged her psychiatric patients and an odd assortment of foreign students to share her quarters. A young woman caretaker helped her with the household.

On Friday nights Dr. Roughton held her weekly soirées, where the notable event would be the serving of the cheese, a massive chunk of Stilton from which issued forth an obnoxious smell. At one of these evenings, Dr. Roughton offered to take me in as one of her resident students. I was happy to escape my cold and unfriendly room, so I promptly moved in, taking a bed in a dormitory-sized room with three other male students. At night, Dr. Roughton, dressed in a leather-and-sheepskin RAF suit and helmet, slept on the veranda outside our room, with its doors wide open, even in the middle of winter. Her loud snoring often kept us awake.

One of the students in our large shared room was a Greek whose distinguishing feature was his black academic gown, which had only a few tatters of cloth left, hanging from his back. He was often in trouble and was fined for not wearing his regulation black gown at the university. Several times he borrowed Professor Roughton's bicycle, which was always parked outside the house, and had to face the fury of the professor, who bicycled to Trinity every day after lunch. This student was also the official photographer of the university newspaper, the *Varsity*, because this gave him opportunities to photograph young ladies privately in the nude.

Due to a shortage of funds and a lack of clothes-washing facilities at Adams Road, the aromas in the room at night where we slept sometimes became overpowering. One of the students was an upper-class Englishman, who had attended Winchester College before coming up to Cambridge. One night when I was retiring and removing my socks, he parodied an incense-bearer, parading around my bed, swinging his arm and chanting in Latin.

An African student doing a Ph.D. in history also lived at the Roughton house. He claimed to be the son of a tribal chief in Nigeria. He was often quite hostile towards the rest of us, and would play the grand piano for hours at night in a large living room adjoining the room where we slept. When I suggested that he cease his piano playing earlier so that we could get some sleep, he rudely dismissed me. This fellow—tall and handsome, with tribal scars on his cheeks—was hugely popular with the young ladies who attended the parties at the Roughton residence.

When my tutor at Trinity heard about my living at 9 Adams Road, he summoned me to his rooms. Wasn't I aware that this notorious house was out of bounds for Cambridge students? This ended my six months of interesting and cheap—indeed, *free*—lodgings at the Roughton residence. On several occasions I had dutifully presented my rent money to Dr. Roughton while she was in the kitchen preparing her Stilton-cheese feasts. But she always refused to take any payment.

I met my future wife, Bridget Flowers, for the first time at one of the weekly dances held at the Cambridge City Hall, and later we coincidentally met at one of the weekly Adams Road cheese soirées. I was pleased to happen upon her sitting on the stairs leading up to Professor Roughton's apartments, nursing a drink. We made a date to meet the next day on the bridge over the River Cam behind Trinity. This began an ongoing relationship. In 1958, my last year at Cambridge, we were married in her hometown of Norwich, in Norfolk.

After the upbraiding by my tutor, I had to search for new accommodations. I located a room in a house on the outskirts of Cambridge. In contrast to the bizarre and lively activities at Adams Road, the new household consisted of a brusque landlady, with many stern rules for her lodgers, and her meek husband. Instead of smelling of Stilton cheese and ripe socks, this house reeked of disinfectant and linoleum-floor polish.

*

My supervision in Hoyle's rooms at St. John's College had become intermittent. Hoyle's supervision of students was generally known to be lackadaisical, as he was completely wrapped up in his own research. Hoyle eventually became famous through his research into the cosmic origin of the elements as well as his steady-state cosmology. In addition to his celebrated work on the origin of the heavier elements in stars, in collaboration with William Fowler and Geoffrey and Margaret Burbidge, he also contributed significantly to explaining how hydrogen, helium and other lighter elements such as lithium were produced in the early universe, thereby indirectly supporting the Big Bang model, which he despised.

One afternoon, Hoyle invited me to his cottage some distance from Cambridge for tea. It was a comfortable house, and his wife greeted me and a divinity student from Trinity in a welcoming way. We had tea and cakes in a sunlit room with a pleasant view of the old-English garden. After tea, we were invited to sit in the living room, where a large oil portrait of Fred Hoyle hanging above the fireplace dominated the room. As in his St. John's College rooms, I couldn't help being distracted by the bold stare from the portrait, while Hoyle himself, staring just as intensely, launched a verbal attack on the divinity student.

"Why do you waste time trying to prove the existence of God?" Hoyle asked abrasively. "Surely an intelligent person like you does not actually entertain notions that there is a God."

The divinity student hesitated and said politely, "The existence of God is a belief, a matter of faith, not proof. It's important for me to sustain this faith."

Hoyle snorted and said, "Faith? Faith in such nonsense?"

The poor divinity student sat quietly, his face beginning to droop and turn pale, as Hoyle became inspired to greater heights in his attack.

Fred Hoyle was a notorious atheist at this time, with no patience for religion. I began to realize that I was witnessing the assassination of a Cambridge divinity student. I wondered whether Hoyle had invited him to tea simply as a form of sport, to exercise his anger at and resentment of religious beliefs, which still dominated many aspects of Cambridge college life. Hoyle's atheism could be seen in his current main research interest—his steady-state model of the universe. In this model, there was no beginning to the universe, and galaxies were spontaneously born in a cold, dark cosmic void, doing away with the need for a creator. Hoyle despised Pope Pius XII's support of George Lemaître's Big Bang theory. To the pope, a beginning of the universe with a dramatic explosion proved the existence of the Creator. Lemaître himself, a priest as well as a cosmologist, eventually managed to dissuade the pope from promulgating these beliefs from the Vatican, for he did not wish to mix science and religion. But to Hoyle, the pope's approval of the Big Bang theory was anathema.

Hoyle never addressed a word to me at the science-versus-religion tea party. I was merely the silent witness to the abusive demise of my fellow student. I was not personally affected by Hoyle's anti-religion diatribe because my parents had rarely taken me to church as a child, and if pressed, I would have described myself as an agnostic. But I felt mortified and sorry for the divinity student. When Hoyle finished his attack, the four of us walked out into the garden, where Mrs. Hoyle showed us her magnificent beds of perennial flowers. As if nothing out of the ordinary had happened, Hoyle then transformed himself back into the convivial country gentleman and college don. There was no further talk about God.

Much later in his life, Hoyle appeared to change his atheistic stance dramatically. In his work on the evolution of stars, Hoyle discovered the need for a special excited energy state, or "resonance," of carbon to get the chemical processes going that would produce the stable carbon nucleus that is the basis of life. Proponents of what is now called the "anthropic principle" turn the process on its head and claim that the existence of carbon-based life (e.g., humans) is proof enough that this rare carbon resonance exists. They also claim that the fundamental constants in nature have their specific values *because* we exist. Hoyle's carbon resonance was discovered experimentally by William Fowler in the mid-1950s. He won the Nobel Prize for the discovery almost thirty years later, along with the theorist Subramanyan Chandrasekhar, while inexplicably Hoyle did not share in the prize.

In the 1980s, Hoyle published popular books on the origin of life, such as *Evolution from Space: A Theory of Cosmic Creationism*. He concluded that it would be almost impossible for life to have evolved through natural processes alone, without the guiding hand of a greater intelligence. The well-known "Hoyle's fallacy," in which he compared the evolutionary origin of life to be as likely as the sudden assembly of a Boeing 747 from a tornado passing through a junkyard, is often used by believers in intelligent design to bolster their position.

*

I felt quite isolated at Cambridge in my research, being the only active relativist there, besides Felix Pirani. Dennis Sciama, who had been one of Professor Paul Dirac's few students, was engaged in interesting research attempting to understand the origin of inertia, the property of a body moving at constant speed until an external force changes its speed. This very basic concept in physics was described in Newton's first law of mechanics. But the origin of inertia had always been a mystery, and had not even been properly understood in Einstein's theory of gravity. Decades later, inertia would prove to be an important area of research in my modified gravity theory (MOG), but in the 1950s, Sciama was as alone in his research focus as I was.

During my second year at Trinity, Roy Kerr, a New Zealander, arrived in Cambridge to begin research in theoretical physics. We became friends, and he asked me for advice on what research topics he should pursue. Roy confessed that his knowledge of physics was limited. In New Zealand he had been working as an applied mathematician, solving differential equations associated with myxomatosis, a disease that was endemic among the rabbit population there. Roy was a brilliant mathematical physicist who had great talent for solving differential equations in applied mathematics. I told him that perhaps he should work on Einstein's gravitation theory. I explained that he didn't need to know much physics, but that there were lots of equations to solve.

I also pointed to the problem of the motion of particles in Einstein's gravity theory, as determined by the non-linear field equations. As Einstein, Infeld and Hoffmann had shown in a celebrated paper published in *Annals of Mathematics* in 1938, the motion of bodies in Einstein's gravity theory was not a separate postulate but could be derived from the theory's basic field equations. In contrast, Maxwell's field equations for the electromagnetic field were linear differential equations, and the motion of electrically charged bodies was determined by an additional postulate called the Lorentz force law. I was working on the problem of the motion of particles in Einstein's theory myself for part of my Ph.D. thesis. In particular, I suggested to Roy that he work on the motion of spinning particles, as the Greek physicist Achilles Papapetrou and others had been doing, as this work looked promising. Roy Kerr's first research project was to work out the static, spherically symmetric solution of my first modified gravity theory, which I had published in the *Proceedings of the Cambridge Philosophical Society*. Indeed, he wrote an excellent paper on this, which was published in

Il Nuovo Cimento. It appeared in 1958 and was titled "On Spherically Symmetric Solutions in Moffat's Unified Field Theory." Roy eventually became famous for discovering the exact, rotating black hole solution in Einstein's gravity theory, one of the few exact solutions of the field equations of the theory. I like to take a small morsel of credit for Kerr's great success.

Later at Cambridge, I had the misfortune of discovering serious mistakes in a paper published by Einstein and Leopold Infeld in 1949 in the *Canadian Journal of Mathematics.* The editors of the journal felt this paper, "On the Motion of Particles in General Relativity Theory," was so important that its first page consisted of the portrait of Einstein taken by the famous Canadian photographer Yousuf Karsh. While studying this paper for my research on the motion of bodies in Einstein's gravity theory, I discovered that there was an error in the derivation of the equations. I asked Roy to check the paper, which he did, and he agreed that there was a serious problem. We decided to write a paper together pointing out this problem. Indeed, we felt obliged to do so for the sake of posterity, alerting researchers to the fact that this famous paper was actually not correct. We submitted our paper to *Physical Review,* the premier physics journal in America.

To our complete dismay, several weeks later we received a letter from Peter Bergman, an editor at *Physical Review* and a former assistant to Einstein at the Institute for Advanced Study in Princeton. The letter stated that *Physical Review* could not publish our paper because it would "besmirch the reputation" of one of the greatest physicists of the twentieth century. I was so upset by this intellectual dishonesty that I contemplated leaving physics. I suffered deeply from an idealistic notion of how science should be conducted, seeking the truth and trying to understand the workings of nature. (Some might consider this *youthful* idealism, but more than fifty years later, I still feel the same.)

The way the leading American journal in physics treated our paper was disillusioning to me because our paper was not wrong; rather, Einstein's reputation was considered more important than the truth. Einstein had died in 1955, so of course I was unable to correspond with him about this. I felt certain that he would not have condoned this censorship by the *Physical Review*. When I had first discovered the error in the paper, I had written to Leopold Infeld, who had left the University of Toronto and was then a professor in Warsaw. He never replied to my letter. I wrote again, pleading for his help, explaining that I was a student at Cambridge finishing my Ph.D., and it was important for me to know what to do about this error, which made the main arguments of the paper false, because the subject of the paper formed part of my thesis. Again, I did not receive a reply.

*

I often went to the weekly informal discussions held in Dennis Sciama's rooms at Trinity, where, in addition to Sciama, Ivor Robinson, a relativity theorist who was visiting Cambridge, and Felix Pirani were sure to be present. On one such occasion, Roger Penrose was also present.

Penrose was a year or two ahead of me at Cambridge, having just finished his Ph.D. in mathematics at St. John's College when I met him. Roger came from the well-known Penrose family. His brother, Oliver, was also a physicist, specializing in condensed matter physics.

When I first made his acquaintance, Roger was deliberating whether his future should be as a mathematician or a theoretical physicist, and we discussed relativity theory. As with Roy Kerr, I suggested to Roger, as did Sciama and Robinson, that he apply his brilliant talent for mathematics to Einstein's gravitation theory. We emphasized to Roger that there was a great need in relativity theory for solutions of mathematical problems in the theory. The physics problems, demanding research in particle physics and quantum field theory, already constituted an active program at Cambridge, spearheaded by Abdus Salam and Paul Matthews.

Penrose was a quiet young man of slight build, with a shock of black hair and a pale, studious face. He appeared to be of a calm disposition and spoke with a soft, upper-class-English voice. Roger took to heart our suggestions about future research on gravitation theory, and as time eventually showed, he did seminal work in relativity theory; for example, inventing the celebrated conformal Penrose diagrams, which had a significant application in black hole physics. In 1965, he published a paper in *Physical Review Letters*, "Gravitational Collapse and Space-Time Singularities," formulating a rigorous theorem about the necessity for an essential singularity occurring at the centre of the Schwarzschild black hole solution of Einstein's field equations. This singularity was associated with the event horizon that formed when a too-massive star collapses under its own gravitational attraction and becomes a black hole. This is a mysterious solution of Einstein's field equations, which began to come into prominence with the work of John Wheeler and his students at Princeton. Later, Penrose collaborated with Stephen Hawking in proving that, given certain assumptions in Einstein's general relativity, a singularity at the time of the Big Bang was inevitable. These published papers were partly responsible for a revival of interest in Einstein's gravitation theory in the late 1960s and 1970s.

*

While I was busy working towards my Ph.D., my mother and father were not doing well in Denmark. My father was still suffering the effects of TB, and my aging mother was working long hours on her feet in restaurants. When Mr. Sanderson, of the Nuffield Foundation in London, heard of my parents' plight, he decided to help out by bringing them to Cambridge. With improved health, he believed my father would find it easier to locate work, which was scarce in Copenhagen at that time, especially for foreigners. He also felt that the stress of worrying about my parents would adversely affect my doctoral research.

My parents arrived in England in 1956, and I first found rooms for them to rent in a house in Cambridge, and then a comfortable house on Oxford Road not far from the colleges. I moved in with them, which proved to be a great help to me, for my mother was able to feed me and keep my clothes laundered so that I could concentrate fully on my research and on my girlfriend, Bridget, who had a room above the George and Dragon Pub in Cambridge.

7

———

THE EINSTEIN FEST

O NE DAY IN JUNE 1955, as I was walking through Trinity Great Court, a senior fellow of the college waved to me across the court and approached me, walking on the sacred college lawn, which, as a fellow, he was permitted to do, and as a lowly student I was not. "Moffat, I've been asked to tell you that we would like you to attend the fifty-year commemoration of Einstein's discovery of special relativity, in Bern, Switzerland. Mr. Hoyle has been invited to attend the commemoration to represent Cambridge, but he is not able to be there."

I knew, of course, that Hoyle was away in the States doing research. Also, Hermann Bondi, a fellow at Trinity College and a collaborator of Hoyle's, had a few months earlier been appointed professor of mathematics at King's College London, and therefore would probably represent King's College rather than Cambridge at the conference.

"I'm only a research student," I answered. "Wouldn't it be more appropriate for a senior person to attend?"

The fellow assumed a stern look and said, "But you're the only person doing work on relativity and gravitation at present, and I've heard you're publishing papers on gravitation in our Cambridge *Proceedings*. Is that right?"

"Yes, that is correct."

"Our college bursar will provide you with 30 pounds to cover your travel expenses."

This amount would barely cover my travel expenses, and anything additional would have to come out of my meagre grant money. However, it seemed inappropriate for me to suggest that this would not be enough expense money, or that I wouldn't like to attend the ceremonies in Bern. Besides, I decided that it might be fun, and I would meet some famous physicists. I agreed to go to Bern and represent Cambridge. I only wished that I could have met Einstein himself there, but he had died only two months before.

In fact, I had written a final letter to Einstein on April 15, 1955. Since he died on April 18, it would not have reached him in time. "Dear Professor Einstein," I had written. "Since I received your last letter two years ago in Copenhagen, I have been fortunate to become established at Trinity College, Cambridge, as a research student. I send you enclosed some of my recent work on generalizing gravitation theory." The paper was my first published generalization of Einstein's gravitation theory, based on a complex symmetric metric tensor and complex Riemannian geometry.

I had been eager to hear what Einstein would make of this new, original modification of his gravity theory. Modifying Einstein's gravity theory as he and I were attempting was not a popular endeavour at the time, and consequently I did not have other people I could discuss this subject with, including my supervisor. When I heard of Einstein's death in the news, I felt very saddened, knowing that the world had lost one of the greatest physicists of all time, and I had also lost the personal connection with this great man who had been instrumental in my entering academic life and pursuing my goal to become a physicist.

*

The following week, I took a train from London, crossed the English Channel by boat and caught a connecting train in Paris to Bern. This was the first time I had travelled in Europe in so grand a style, and I was enjoying the new experience. I arrived in Bern a couple of days before the beginning of the conference, to acclimatize myself to the city. I went to see the famous Zytglogge, the clock tower in the heart of the old town. It was a bright, sunny day, and the historic Swiss buildings with their intricate architectural details impressed me.

As I stood in the clock-tower square, a red sports car zoomed in, and someone who looked like a clone of the Hollywood film star Cary Grant got out, flashing a grin. He asked, in a strong American accent, "Are you a physicist attending the meeting?"

"Yes, I am," I answered. "How did you know I'm a physicist?"

"Well, there are ways of figuring this out," he said, laughing. He shook my hand. "I'm Stanley Deser," he said. "I'm on sabbatical leave in Europe at the moment, and dropped in to attend this meeting. I guess it should be interesting. Where are you from?" I explained that I was representing Cambridge.

Stanley flashed his smile at me again and said, "We'll see each other again at the meeting in a couple of days. Hope you have a good time in Bern." He hopped into his red sports car and zoomed off.

I had heard of Stanley Deser in connection with his work on relativity theory. He went on to do seminal work in gravitation theory at Brandeis University, in Massachusetts. With collaborators, he developed a fundamental way of solving Einstein's gravitational equations.

The next day, during another perambulation through Bern, I ran into Bill Bonnor, who had been one of the professors that Greenall, at the Department of Scientific and Industrial Research, in London, had sent me to visit a year before. Bonnor was walking towards me on a charming, narrow street, accompanied by Geoffrey Stephenson, a mathematical physicist at Imperial College London. Bonnor smiled amiably at me and said, "Well, I understand that you are now a student at Trinity College. And here you are at this meeting. How did that come about?"

I told him about standing in for Fred Hoyle. "At the moment I am the only one publishing gravitational research at Cambridge," I added.

Stephenson said, "I gather Felix Pirani is here, but I hear that he is going to King's College in London, so I guess officially you're the one they chose to represent Cambridge."

The evening before the beginning of the conference, there was a reception at the University of Bern. The chief organizer of the conference was Professor André Mercier, a senior professor in the Department of Physics at Bern. The number of delegates at the meeting was small, compared to attendance at the large conferences that now take place in relativity and gravitation, which can number in the several hundreds. Among the famous physicists I shook hands with that evening was Wolfgang Pauli from Zurich, a short, stocky man with a big head and a rosy complexion. Pauli was one of the founders of quantum mechanics and went on to make important contributions to quantum field theory.

I also met Professor Satyendra Nath Bose, the Indian physicist famous for developing "Bose statistics" for what were later called "bosons," which are particles with integer quantum spin such as the photon.* Later, in quantum physics, bosons became understood as

---

*Like spinning tops, elementary particles carry a quantity known as angular momentum or spin. Unlike spinning tops, the angular momentum of elementary particles comes in discrete quantities. Elementary particles fall into two groups, with spin represented either by integers 0,1,2, or half-integers such as 1/2 or 3/2. The integer spin particles such as the photon are called bosons (named after Bose) and the half-integer particles such as the electron and proton are called fermions (named after the Italian physicist Enrico Fermi). Wolfgang Pauli worked out the properties of spin in 1927. He introduced the exclusion principle, which prevents two electrons with the same spin from occupying the same quantum state in an atom, and which provides a quantum mechanical explanation of the periodic table of elements.

the carriers of forces in subatomic physics. The important feature Bose discovered in his statistics is that the photons (bosons) are identical particles, whereas physicists had assumed that there were different kinds of photons. The Bose statistics were later called "Bose-Einstein statistics" because Einstein promoted Bose and made an important contribution to the subject. Bose looked dashing in his black beret and dark suit that evening.

I also shook hands with the celebrated German mathematical physicist Hermann Weyl, who was famous for his contributions to quantum mechanics and was the inventor of gauge theory, which would play a fundamental role in modern quantum field theory. I had read his book, *Space, Time, Matter*, which was one of the first books describing Einstein's general relativity. He seemed like an amiable, kind man, and he was interested in discussing my research. While we were talking, the American physicist Robert Oppenheimer strolled over, a drink in his hand, and Weyl introduced us. I was struck by Oppenheimer's intense, light blue eyes and his handsome, aquiline face. I later observed that he always wore a three-piece grey suit with a shirt and tie, and looked more like a wealthy banker than a physicist, although in those days, in contrast to today, well-known physicists were usually attired in formal suits rather than T-shirts and jeans. As I stood hobnobbing with these famous physicists, it dawned on me that I was the youngest participant at the conference—just a second-year research student, only twenty-two years old.

We were gathered at this meeting to commemorate Einstein's revolutionary paper on special relativity of 1905, "On the Electrodynamics of Moving Bodies," which changed our understanding of space and time forever in physics. In addition, there would be lectures on the developments of Einstein's general relativity. There were also lectures on Einstein's attempts to unify gravity and electromagnetism by, for example, Bruria Kaufman, an Israeli physicist who had been Einstein's chief assistant in his later years at the Institute for Advanced Study at Princeton. She had devoted much of her research to investigating Einstein's nonsymmetric field theory generalization of general relativity.

There was a session, too, on the significance of the gravitational waves prediction in Einstein's gravity theory. At the time, there was much controversy about whether gravitational waves existed or not. Indeed, in 1936, Einstein and his assistant Nathan Rosen had published a paper on the properties of cylindrical gravitational wave solutions of Einstein's field equations. The title of the original version of the paper was, "Do Gravitational Waves Exist?" This version was rejected by *Physical Review*. The anonymous referee of the paper was eventually revealed to be a young Caltech assistant professor, Howard Percy Robertson. This was the first paper by Einstein that was sent to an anonymous peer reviewer, and it was the first paper that Einstein had had rejected. In fact, it is not clear whether Robertson actually rejected the paper, or whether Einstein jumped to conclusions, being unfamiliar with the peer review system. Einstein probably resented the anonymous reviewer proposing modifications to the paper and claiming that there were errors in it. He reacted by treating the reviewer's criticisms as an outright rejection of the paper.

Years later, I met Nathan Rosen, Einstein's collaborator on that paper, at a bus stop in Padova, Italy, while attending a conference on general relativity. We fell into conversation and I asked him about this paper. Rosen told me that Einstein had been angry at the rejection. Rosen had been present in Einstein's office that particular morning when his secretary brought in the mail with the letter from *Physical Review*. Upon reading the letter, Rosen said, Einstein leapt out of his chair and threw the envelope with the letter and the manuscript into his trash can, which he then kicked loudly around his office. He promptly wrote to the editor of *Physical Review* vowing never to submit a paper to the journal again. In the letter, he criticized the editor for not warning him upon submitting the paper that it would be subjected to anonymous refereeing. Fortunately, Rosen told me, he managed to retrieve the manuscript from the trash can, unbeknownst to Einstein. He revised it in accordance with the referee's criticisms, and submitted it to the *Proceedings of the Franklin Institute* in Philadelphia, a journal that published manuscripts on entomology and various other specialized topics, not necessarily related to physics. They, of course, were delighted to receive a paper from Albert Einstein, one of the most famous physicists in the world, and they published it without peer review.

Apart from one letter published in the *Physical Review*, answering criticisms put forth by a chemist-physicist on Einstein's non-symmetric unified field theory, Einstein was true to his word to the editor of the journal. The paper with Rosen on gravitational waves was probably the only genuine encounter Einstein had with anonymous peer review. In the early twentieth century, it was not common for German physics journals to send out papers for anonymous review, as is the case today. Normally, the editor of the journal would make the decisions as to whether papers would be published or not. As Einstein's fame grew, his papers were almost automatically accepted for publication. Significantly, his famous five papers from 1905, as an unknown Swiss patent official, were accepted by the editor of the prestigious German physics journal *Annalen der Physik* without any peer review.*

---

*In addition to "On the Electrodynamics of Moving Bodies," which introduced special relativity, the other papers Einstein published in his "miraculous year" of 1905 were "Does the Inertia of a Body Depend on Its Energy Content?," which discussed the idea of the equivalence of mass and energy and contained the famous equation $E=mc^2$; "On a Heuristic Point of View Concerning the Production and Transformation of Light," which asserted that light occurs not only in waves but in particles and led to the development of quantum physics; "On the Movement of Small Particles

After Einstein relocated to Princeton in 1933, he began to publish in American journals in English. The *Physical Review* had taken on the mantle of the leading journal in physics in the United States. Its editor at the time was John Tate. Einstein published two papers in the *Physical Review* before the trash can incident. The first was the famous 1935 paper by Einstein, Boris Podolsky and Nathan Rosen on what came to be known as the Einstein-Podolsky-Rosen (EPR) Paradox, which became one of the most cited papers in modern physics: "Can Quantum-Mechanical Description of Physical Reality Be Considered Complete?" This paradox was one of Einstein's thought experiments that he promoted to criticize quantum mechanics and was part of the celebrated Bohr-Einstein debate that took place over several years in the 1930s and 1940s. Einstein, Podolsky and Rosen claimed that quantum mechanics was an incomplete description of nature. In fact, the physical interpretation of quantum mechanics remains a controversial issue today.

The second and last paper Einstein published in *Physical Review* was later in 1935, and also in collaboration with Rosen. The subject of "The Particle Problem in the General Theory of Relativity" was the Einstein-Rosen Bridge, known today as the "wormhole" solution of general relativity. Beloved of science-fiction writers, the wormhole was Einstein's attempt to remove the problem of unphysical singularities in his gravitation theory. The wormhole is a mathematical portal in spacetime, allowing a space traveller to move more or less instantaneously through the universe and come out in a distant part of it, or into another universe.

One wonders whether Einstein could ever have succeeded in publishing many of his iconoclastic papers had he been subjected

---

Suspended in Stationary Liquids Required by the Molecular-Kinetic Theory of Heat," on Brownian motion; and "A New Determination of Molecular Dimensions," based on Einstein's Ph.D. dissertation, which calculated Avogadro's number and the size of molecules.

to the draconian type of peer review that prevails in the physics establishment of today.

*

Einstein's problematic paper "Do Gravitational Waves Exist?" proved to be a major topic at the Bern Einstein Fest. It was compelling to think that gravitational waves, like electromagnetic waves, existed. Yet, in 1936, Einstein wrote to his colleague Max Born, "Together with a young collaborator [Rosen], I arrived at the interesting result that gravitational waves do not exist, though they had been assumed a certainty to the first approximation. This shows that the non-linear general relativistic field equations can tell us more ..." In his paper with Rosen on gravitational waves, certain mathematical mistakes were made, and Robertson criticized these mistakes. In the published paper, the mistakes were corrected, and it was not stated categorically that gravitational waves did not exist. Indeed, at the time, Einstein vacillated between accepting the existence of gravitational waves and not accepting them.

Another assistant of Einstein's, Leopold Infeld, who did not believe that gravitational waves were a physical prediction of Einstein's equations, and were not realized in nature, gave a talk at the Bern conference.* Vladimir Fock, who was famous for his work on quantum field theory, having invented what is called the "Fock space," also talked at the Bern conference. In contrast to Infeld, he fervently believed in the existence of gravitational waves.

Infeld and Fock gave talks one after the other at the meeting. Fock, a large, stocky Russian who kept adjusting his hearing aids since he was completely deaf without them, gave the first talk, presenting his reasons for believing in the gravitational wave solutions in

---

*I would write to Leopold Infeld two years later, asking for information about the mistake I had found in his 1949 paper with Einstein.

Einstein's theory. Infeld stood off to one side near the podium as Fock spoke. He was also a large man, a Polish physicist from the University of Warsaw. He was already beginning to bristle with indignation over Fock's talk. When Fock finished speaking, Infeld went to the podium. When he started speaking, presenting his reasons for not believing in the existence of gravitational waves, Fock ostentatiously removed his hearing aids, raising laughter in the auditorium. Even Hermann Weyl, the chair of the session, smiled with amusement.

One of the presentations at the gravitational waves session was given by Hermann Bondi, who, in collaboration with his group at King's College London, reported on new attempts to develop a more rigorous foundation for understanding the role of gravitational waves in Einstein's non-linear gravitation theory.

At another session, chaired by Robert Oppenheimer, a young professor from an American university presented his recent work on relativity theory. Oppenheimer suddenly displayed an aggressive aspect of his personality. He interrupted the speaker and criticized him scathingly in a loud voice. The younger physicist went red in the face and seemed to have the wind blown out of his sails, and continued his talk rather lamely.

*

One afternoon after a session, Professor Mercier announced that we were all to proceed to the university's Department of Zoology for a reception. We assembled in a large hall there, where tables had been set up with food and drinks, which some of the delegates, including Pauli and Bose, appeared to imbibe too heartily. After a couple of hours, the noise level in the hall increased significantly, and Pauli's raucous laughter could be heard all over the room. I knew there was going to be another session starting up after the reception, and I wondered how the organizers could possibly think of having another session after all this alcohol had been drunk.

We all streamed back to the physics auditorium, and some of the older delegates like Pauli were visibly intoxicated. The Greek relativist Achilles Papapetrou gave the first talk in the session, describing his work on the motion of bodies in Einstein's general relativity, and I listened attentively because this was a subject that I had begun working on at Cambridge and that would be part of my Ph.D. thesis. Pauli sat in the second row, nodding his head in a nervous tic.

When Papapetrou finished speaking, he descended from the podium and made the mistake of standing near Pauli. Someone asked Papapetrou a question, and he answered it by quoting Pauli on some technical matter having to do with the motion of bodies. Papapetrou was a tall man with a long neck and an Adam's apple that thrust up and down as he spoke. Pauli looked up at Papapetrou and shouted at him, "Papapetrou, you are a Papageno!" in an insulting allusion to the character in Mozart's *Magic Flute.* "If you have to quote me, Papageno, do it correctly or not at all! I refuse to have such nonsense attributed to me!" Papapetrou's face reddened; he looked mortified. This was the first time I witnessed the famous Pauli publicly humiliating another physicist, his tongue no doubt sharpened by alcohol.

One day, the conference organizers took the delegates on a trip to Lake Brienz in the Bernese Mountains. As we sailed around the lake, I had the opportunity to stand next to Infeld and Fock, who continued to attack one another on the issue of gravitational waves. This time Fock did not remove his hearing aids.

We were let off at a small port on the lake, and Bonnor, Stephenson and I took a walk and viewed the magnificent mountains above us and the beautiful deep blue lake shimmering in the sun beneath us. Sure enough, as we walked back down into the village where the boat was docked, we heard Pauli laughing and shouting loudly from inside an inn, as he led another drinking session. Young and innocent as I was then, I was taken aback by the drunken, argumentative behaviour of some of the older delegates.

*

I don't know whether those older delegates thought much about Einstein during that week in Bern—perhaps they were too jaded at that point in their lives—but I certainly experienced feelings of sadness and regret that Einstein had died just two months before this meeting, while my final letter to him was on its way. I walked by the building where his apartment had been located, at 49 Kramgasse, where he had lived with his first wife, Mileva, and their infant son, Hans Albert. I walked over the Kirchenfeld Bridge that he had crossed every day to go to the patent office, where he worked as a second-class civil servant.

I wondered whether before he died, Einstein had received an invitation to this important Einstein Fest celebrating the fiftieth anniversary of his discovery of relativity, and whether he had considered attending. Einstein had never returned to Europe after fleeing Nazi Germany and taking up his position at the Institute for Advanced Study in Princeton in 1933, so perhaps he would have declined this invitation as he had so many others. I felt that it was a pity for me, personally, that he had not been present during that week in Bern, because I would have been thrilled to meet him in person, and complete our cycle of correspondence with a personal conversation about physics.

8

———

WOLFGANG PAULI

*I*N THE LATE FALL of 1955, my second year as a Ph.D. student at Trinity College, our student supervisor, James Hamilton, announced that Wolfgang Pauli would be visiting Cambridge for a couple of days. With Pauli's rude behaviour at the Bern conference still fresh in my mind, I reacted to this announcement with a mixture of excitement and dread.

On the morning of Pauli's arrival, I hurriedly washed up the breakfast dishes, left my room earlier than usual and walked quickly over to the Arts School, checking in at the local Barclays Bank on my way, to see how overdrawn my account was. I was excited about Pauli's visit because he had been working recently on the same problem that had been occupying me, and I was hoping to get some insights from him, although I doubted that he would be interested in the opinions of a second-year research student. This is where the dread entered: after having observed Pauli's behaviour in Bern, I wondered what might go awry during his visit to Cambridge.

On reaching the main room on the second floor of the Arts School, I encountered our supervisor. James Hamilton was a senior lecturer at Cambridge who specialized in particle physics and had published a textbook on the subject. He was a tall, severe-looking

Irishman with cold cornflower-blue eyes and a fine, regular-featured face with greying hair at the temples. Jim had served in the British army during the Burma campaign of the early 1940s, and suffered occasionally from bouts of malaria. He was known for his acerbic comments and authoritative manner, and was feared by most of the research students.

"Ah, Moffat," he said, giving me his habitual icy-blue stare. "Our distinguished guest will be here shortly. I would like you to take Professor Pauli to lunch. Your colleague Ian McCarthy has agreed to go with you."

This was a disturbing request, further complicating my feelings about his visit. Pauli was a winner of the Nobel Prize, awarded to him in 1945 for his discovery of the exclusion principle—postulating that two electrons could not occupy the same quantum state simultaneously in an atom—which played a fundamental role in the development of quantum mechanics and atomic physics. He had also been instrumental in developing quantum field theory in collaboration with Werner Heisenberg and Victor Weisskopf. How could Hamilton expect two research students to take the great Pauli to lunch? "Sir, perhaps a senior member of the department should have this honour," I suggested.

Hamilton straightened up and smoothed back his hair with a nervous thrust of his hand. He cleared his throat. "Nobody is available to take him to lunch today, so you'll have to make the best of it. Are you refusing to do this?"

I felt the tension rising. "No, sir, I'll go with Ian," I said. "If he's agreed to do this, well, then I'm ready to help him."

Hamilton shoved his hand in his pocket and drew out a ten-pound note. "This should cover the lunch, but be sure to get a copy of the bill and give me back any money left over."

"Yes, sir."

"Pauli will be here at about noon, so wait here for Ian McCarthy," Hamilton said. "Pauli was told to come to the Arts School."

With this command, Hamilton left. I heard him walk heavily down the stairs and slam the door as he exited the Arts School.

*

Ian arrived about thirty minutes later, grinned at me and said in his broad Australian accent, "Well, John, I hear we're taking the great man to lunch."

"The two of us taking Pauli to lunch!" I exclaimed. "What about Dirac or Hamilton himself, or one of the other senior people?" I stuck my hands in my pockets and said, "Hamilton gave me ten pounds to pay for lunch. I hope that's enough. I'm overdrawn in my bank account."

"So am I," Ian said, and his grin broadened.

Ian was studying nuclear physics. He was from Adelaide, Australia, and was a year ahead of me in his Ph.D. research. I had come to know Ian well thanks to an old sports car I had bought, a red Singer Le Mans two-seater that had supposedly raced in the famous French Le Mans car race in 1932. I had swapped it for my BSA motorcycle and a few pounds. I had never driven a car before, and eventually this one would cause me a lot of heartache and expense. On the day it was delivered, I unwisely started it up and put it into first gear. The little red car promptly raced down a hill and crashed into a fence. Fortunately, my foot managed to find the brake pedal in time, and the front bumper was only scraped.

When I told Ian about this incident the next day, he laughed and offered to help me out with my new red sports car. He taught me how to drive the wretched car; we could often be seen roaring around Cambridge with me trying desperately to change gears by a double-declutch maneuvre, since the cars of the early 1930s did not have a modern clutch system. Through the driving lessons, Ian and I became close friends. He often invited me home to his flat to have dinner with him and his wife, who was also from Adelaide.

The red car had recently come to a bad end. One morning when I was driving to town, the crankshaft shot its way through the engine like a torpedo, and the car came to an abrupt halt. I shot out over the open windshield, but fortunately wasn't seriously hurt. I could not afford to repair this damage, and that was the end of the little red sports car.

Ian and I waited for Pauli in a room on the second floor of the Arts School where the theoretical physics research students met to talk about cricket matches, arrange tennis games and, occasionally, discuss physics. At one o'clock there was a sudden loud banging on the door. I jumped up and opened it, to face the familiar short, obese figure of Pauli in a wrinkled dark suit and a tie that looked twisted out of shape. Pauli stared at me with his protruding brown eyes and said, "*Guten Tag!* Are you the student who takes me to lunch?" Obviously he had no memory of me from the summer conference in Bern.

"Yes," I answered nervously. "I'm John Moffat, and this is my colleague, Ian McCarthy. Dr. Hamilton asked us to take you to lunch and perhaps show you around Cambridge."

He shook my extended hand vigorously and then marched into the room and shook hands with Ian. "*Ach*, well, let us go to lunch, I'm hungry," he said peremptorily. "I do not wish to be shown around Cambridge. I have been here before and have seen the colleges and the town. Once is enough! Hamilton told me that you students will entertain me after lunch. In this room, I believe."

This was the first I had heard of this event, so I swallowed hard and said, "Yes, Professor, we'll bring you back here after lunch so that we can discuss physics."

"Discuss physics!" Pauli said harshly. "You will speak to me about your research!"

"Yes, of course, Professor."

Pauli followed us out into Benet Street with its narrow pavements, and we made our way to the King's restaurant on King's Parade. I had never been able to afford a meal there myself, but had often stared through the window when passing by at the dinner guests inside, enjoying a meal with wine. The maître d' showed us to a table next to the window and gave us menus. Pauli did not open his, but glared at the waiter and commanded, "Waiter, bring red wine!"

Some time passed while Pauli sat muttering to himself, and we waited awkwardly for the wine to arrive. Pauli had a large head with thinning dark brown hair, a reddish complexion, and his fat neck was constricted by his shirt collar and the unruly tie. As I had noticed in Bern, he had a nervous tic that took the form of a nodding head. To avoid embarrassment, I stared out of the window at students and townspeople passing by in the fall sunlight. The waiter arrived with a bottle of red wine and displayed the label to Pauli. He grunted his assent and the waiter poured wine into his glass. He came round the table to pour wine into our glasses too, but Pauli exclaimed, "*Nein! Nein!* Put the wine here next to me." The waiter looked uncomprehendingly at me, so I nodded my head in agreement, and he set the bottle on the table next to Pauli.

We ordered lunch, and while we waited for the food to arrive, Pauli drank more than half the bottle of wine. After the first course of soup, he finished the rest of the bottle with rapid gulps from his glass. He eyed me with an authoritative look. "Well, order another bottle, young man!"

"Yes, Professor," I said with dismay as I tried to catch the eye of the waiter.

Another bottle of red wine arrived and Pauli told the waiter to give us a glass each, from which I understood that he intended to finish off the rest of the wine himself. I began to feel concerned about how the rest of the afternoon would unfold. I hoped that Hamilton was not going to accuse me of getting the professor intoxicated before the afternoon's entertainment commenced. After we had finished eating and Pauli had drunk the rest of the second bottle of wine, he became noticeably more voluble. He turned his attention to Ian. "So, young man, where are you from?"

"Australia . . . Adelaide, I'm doing research in nuclear physics."

"Ah, Australia!" Pauli exclaimed. "You must know H.S. Green. He works with Max Born in Edinburgh."

"No, I don't know him personally, but of course I've heard of him and his published work," Ian replied.

Pauli exploded, saying in a loud voice, "He is a *dumkopf*! An idiot, you understand."

"Oh, I see," Ian responded nervously.

I noticed that when Pauli spoke angrily, his head stopped nodding. "I have told Max to get rid of him, but he doesn't listen to what I say these days. This work Green has done on para-statistics in field theory is utter nonsense! What do you have to say about that?" He glared at Ian. I was hoping that Pauli was not considering ordering yet another bottle of wine.

I could sense that Pauli was not going to let Ian off the hook easily. He fixed Ian with a steely-eyed gaze and said, "Are all of you Australian physicists stupid?"

Ian looked at Pauli, speechless. I could see that he was making every effort to control himself. Pauli went back to nodding his head, and stared morosely out the window at the view of King's College that could just be glimpsed in the distance. I thought that it was best to make our way back to the Arts School, to cut off this harangue against Green and Australian physicists. But Pauli stopped nodding his huge head and said, "Now, Abdus Salam, he is here in Cambridge at St. John's College, yes?" I nodded in agreement. "He is brilliant, you understand. He has a great future in physics." He smiled. His face was flushed and his eyes looked somewhat unfocused.

I waved to the waiter and asked him to bring the bill. I paid, pleased that despite the two bottles of wine, the total was less than ten pounds. We left the restaurant and escorted Pauli back to the Arts School and whatever calamitous events awaited us there. As we turned into Benet Street, I took in the variegated scene of King's Parade. Beyond the street lamps King's Chapel rose towards the sky, its pinnacles dark grey against the cumulus clouds. Down at the other end of King's Parade, the sandy-grey bulk of the Senate House jutted out and merged with the narrow beginning of Trinity Street. As we approached the iron gate of the Arts School, a crowd of students poured out, wearing their flowing black academic gowns. Their excited voices echoed in the small confines of the Arts School entrance. I suddenly felt a thrill of anticipation mingled with dread at the prospect of giving an unprepared talk to Wolfgang Pauli.

*

We returned to the research students' room, which was long and narrow with dark oak-panelled walls and a blackboard at one end. A single chair had been placed in the middle of the room facing the blackboard. Paul Dirac was waiting for us, together with Jim Hamilton and several other research students who were seated in chairs against the walls. The room had the eerie atmosphere of a Spanish inquisition. Dirac came forward and shook Pauli's hand, and Pauli smiled with pleasure. I hoped that Dirac was not able to detect the smell of the wine that had been drunk at lunch. I was impressed by Pauli's ability to hold what seemed to my youthful inexperience a large quantity of alcohol.

After some preliminary polite conversation between Pauli and Dirac, Pauli began pacing back and forth across the room, while Dirac stood watching him, his spare, wiry figure in a grey tweed suit smudged by chalk marks, as if he had just given a lecture and hadn't bothered to brush off the chalk. "I have been reading your recent work on quantum field theory, Wolfgang," Dirac said. Pauli stopped his pacing, his head nodding with his nervous tic.

"Yes, Paul, I am still not happy with the lack of rigour of the subject," Pauli responded. "I feel that renormalization theory is just a passing preliminary solution. There is also Werner Heisenberg's unified field theory. This theory of his could unite the forces and explain many mysteries, although I have serious problems with it."*

Pauli looked expectantly at Dirac, who replied, "I agree, of course, that the renormalization program is unsatisfactory. I do not think this cancellation of infinities makes sense. I cannot accept it as a final description of nature. As for Werner's unified theory, I do not think that we should try to solve all the problems at once. We should solve one problem at a time."

Pauli frowned and said, "Well, Paul, the renormalization program does give answers that agree remarkably well with experiment. However, I agree that it cannot be the final answer. As for Werner's unified theory equations, I feel that we must unify the forces of nature and then the correct theory will solve many problems simultaneously."

---

*Calculations of quantum processes, such as the scattering of one elementary particle after colliding with another, lead to infinite quantities. In a renormalizable theory, a mathematical technique can deal with these infinities in such a way that all computational results that correspond to measured quantities yield finite values. Attempts to quantize the gravitational field do not allow for such a renormalization technique to be applied, and therefore unrenormalizable quantum gravity theories do not have meaningful predictions.

"No, no! I must disagree," Dirac responded immediately. "This is not the way to approach theoretical problems. We solve one problem at a time . . ."

Pauli looked irritated at Dirac's response. "No, Paul, physics is solved by unifying principles and this leads us to the correct theory."

Dirac smiled, and said, "Well, Wolfgang, it seems that we have always disagreed on this issue. I am not impressed with Albert's attempts to construct unified field theories. He seems to have wasted years looking for this theory with poor results."

"*Ach!* Albert and his unified theory!" Pauli exclaimed. "Absolute nonsense. He never listened to me. The old man wasted his time!"

I was amused by this exchange between the two great men. It was unusual for us students to witness Dirac verbalize any arguments at such length. In addition, I was disconcerted by Pauli's attitude towards Einstein, which was as dismissive as Bohr's and Schrödinger's had been.

Dirac bid his farewells, as he was on his way to attend a committee meeting. Pauli settled himself into the chair in front of the blackboard and said, "*Achtung*, let us begin!" As his massive head began nodding, the nervous tension in the room became palpable.

Hamilton signalled to Riazuddin to begin. He was a second-year research student working on problems in quantum field theory. He was very quiet and hardly ever spoke to any of us; in his muted behaviour, he was the Pakistani equivalent of Dirac. He had acquired a reputation among us for performing very long, arduous calculations in quantum field theory. He approached the blackboard with chalk in hand and began in tiny, spidery handwriting to carry out one of his lengthy calculations. He had reached the bottom of the blackboard, working in the conventional left-to-right fashion, without uttering a word. Suddenly, Pauli's head stopped nodding and he shouted, "This man is deaf and dumb! I will not witness any more of this uninteresting calculation! *'raus!*" He gesticulated with both arms and poor Riazuddin smiled apologetically and returned to his chair.

Hamilton then signalled to David Candlin, a young research fellow at one of the colleges. He got out of his chair and smiled nervously at Pauli. David was tall and gangly and his face tended to become pink when he was nervous, which he indeed appeared to be now. He also suffered from a slight speech impediment and would stutter when excited. Pauli's head was nodding again. As before at lunch, I realized that Pauli's nervous tic would always stop just as he was preparing to unleash one of his violent verbal attacks. I waited in suspense as David began to speak. He was working on an obscure problem in scattering theory and quantum field theory, which none of us had yet been able to follow. He stuttered and spoke very fast and occasionally wrote some equations on the blackboard. We waited. After about ten minutes, Pauli's head stopped nodding and I knew David was in for it.

"This young man hasn't said a comprehensible word since he started. Utter rubbish! I won't listen to this anymore!" Pauli shouted in his thick Viennese accent.

David lost the chalk on the floor; his face was flushed and perspiration was shining on his brow. Hamilton said, "All right, David, you can sit down now." We waited with apprehension to see who the next victim would be. Hamilton looked at me with an expressionless face, and said, "John, you're next. Don't make your presentation too long."

As I rose from my chair and walked to the blackboard, I felt like Danton approaching the guillotine as Madame Defarge sat knitting in the Place de La Concorde. I began with some introductory comments about Haag's theorem in quantum field theory. This was the same Haag I had met a couple of years before at the Niels Bohr Institute, when I had given my first talk ever, before I was even technically a student. Pauli himself had been working on Haag's theorem and its consequences for quantum field theory, so I knew that I was intentionally entering the lion's den. I soon became caught up in my arguments, however, and temporarily forgot that Pauli was sitting there in his chair like a large toad, his head nodding vigorously. But after about ten minutes, out of the corner of my eye, I suddenly saw that Pauli's head had stopped nodding. I froze.

"Moffat, this is utter nonsense you are saying!" Pauli roared, waving an arm furiously.

At this point I had built up a considerable amount of adrenaline, and I reacted without careful consideration. I shouted back at Pauli, "Professor Pauli, it is not nonsense!"

There was a long moment of hushed silence in the room, and Hamilton looked at me, startled and alarmed. Suddenly, Pauli rose from his chair and moved his bulk towards me where I still stood frozen at the blackboard. He came up close to me, threw his fat arms around my chest and shoulders in a bearish embrace and said, "*Wunderbar!* This young man speaks back!"

My fellow students were in awe of my audacity in disagreeing with the great man. I didn't explain to them that it was simply an ill-motivated knee-jerk reaction due to my nervous state, and that I wasn't sure at all why I had disagreed with him. I did learn from this episode an important lesson that has always remained with me: When you believe that you are right, defend your point of view.

*

Pauli was intensely devoted to physics. His attacks were not personal. He was used to being correct in his intuitive feelings about physics, and he felt impatient with any display of sloppy thinking or carelessness. I realized that this was a hallmark of many of the great physicists of his generation, who had developed the remarkable, innovative theory of quantum mechanics.

In spite of this, it was common knowledge among physicists that Pauli had caused several prominent physicists to abandon important ideas because they had not been able to withstand his withering criticisms. One was Ralph Kronig, a young Columbia University Ph.D. who had studied in Europe for two years. He had discovered the idea that the electron had a quantum spin, and he had proposed the idea to Pauli, who had ridiculed it, saying that it was a clever idea but it had nothing to do with reality. Consequently, Kronig did not pursue this idea and publish it. A few months later, two of the Dutch physicist Paul Ehrenfest's students, Samuel Goudsmit and George Uhlenbeck, suggested the idea of a quantum spinning electron. Another famous Dutch physicist, Hendrik Lorentz, pointed out that the idea of a spinning electron was incompatible with classical electrodynamics. Nevertheless, Ehrenfest insisted that Uhlenbeck submit the paper, and it was published in 1925. It was an important discovery in the development of quantum mechanics, and Pauli, in the year following the paper's publication, successfully included the idea in the formalism of quantum mechanics.

After the embrace at the blackboard, Pauli did not return to his chair. We understood that the entertainment session with the great man was over. Pauli shook Jim Hamilton's hand, turned his big head to me and smiled. "This young man—Moffat—can walk with me and help me find my hotel again. I must rest."

Pauli and I left the Arts School and proceeded back down King's Parade. In spite of his obesity, Pauli walked rapidly beside me down narrow Trinity Street as we headed towards the Blue Boar Hotel opposite Trinity College. "You must work harder at understanding these problems in quantum field theory," Pauli advised me, wheezing slightly from the exertion of the brisk walk.

"Yes, Professor, I will make more efforts to find a way to solve the problem," I said. "I see that you don't agree with Professor Dirac's approach to solving physics problems."

"Absolutely not!" Pauli replied emphatically. "Paul and I have always disagreed about this issue. Ultimately we must unify the laws of nature. I believe that during the next fifty years, physicists will see this as their ultimate goal. Maybe they will even succeed, who knows?"

Today, almost six decades later, unifying the four known forces in nature—gravitation, electromagnetism, and the strong and weak nuclear forces—is as illusory a goal as it was then.

At the door of the hotel, I shook Pauli's hand and wished him a safe journey back to Zurich.

9

———

PAUL DIRAC

I N T H E L A T E 1950s and the 1960s, most theoretical physicists
were working on particle physics and quantum field theory,
to understand the workings of nature at the subatomic level.
This endeavour was boosted by the large amount of data coming
out of the new accelerators at CERN, Geneva and Dubna near
Moscow. In 1956–57, when I had partially completed my Ph.D.
thesis, and felt more confident about where I was headed, I, too,
turned from gravity to these two popular topics and joined the informal particle theory group at Cambridge. I wanted to be involved in
the exciting activities of my colleagues.

In his position as Lucasian Professor of Mathematics, the chair
filled originally by Isaac Newton, Paul Dirac was the senior professor of theoretical physics at Cambridge. His office was in the Mathematics Department at the Arts School on Benet Street, where the
Theoretical Physics Department was also housed. The Cavendish
Laboratory, where much of the experimental physics took place,
could be reached from the Arts School by an alley. Dirac was titu-
lar head of the theory group in physics, which consisted of some
senior lecturers and the research students. He appeared every

Thursday for our theoretical physics seminars. Otherwise, he did not show up very often.

Paul Adrian Maurice Dirac was born in 1902 in Bristol, England, where my parents and I had suffered through such intense Nazi bombings during the war. A fellow pupil at the school Dirac attended there was Archibald Leach, who later became the Hollywood film star Cary Grant. An abstract sculpture in Bristol today celebrates Dirac as a famous native son of the city, and not far away is a lifelike bronze statue of Cary Grant.

Dirac took an engineering degree, but later switched to theoretical physics and became a student and later a fellow of St. John's College, Cambridge. He was one of the co-founders of quantum mechanics, and in 1928, when he was a twenty-five-year-old research student at St. John's, he published his famous Dirac equation describing the spinning electron as both a particle and a wave. The equation also predicted the existence of anti-matter, in the form of the positively charged electron. There was opposition to Dirac's paper from such physicists as Bohr and Oppenheimer. Yet experiment vindicated Dirac in 1932, when Carl Anderson at Caltech discovered experimentally an electron with a positive electric charge. He christened this anti-matter particle the "positron." Dirac and Schrödinger shared the Nobel Prize for physics in 1933, the year after I was born, for their contributions to developing quantum mechanics.

The Dirac equation is one of the most significant discoveries in the history of physics because it predicted a new particle and the existence of anti-matter. At Cambridge, Dirac was a famous but reclusive figure. Because of his extreme reticence and private nature, he was not well known to the public as Einstein was, but behind the scenes he was revered by physicists.

Dirac was not seen much at the Arts School when I was at Cambridge, and much of the supervision of the students was performed by James Hamilton. As a student, I pondered how Hamilton reacted psychologically to doing research in particle physics in the shadow of Dirac's genius and fame. Hamilton seemed to regard me as an odd student due to my anomalous academic background, but was pleasant enough to me most of the time.

Hamilton conducted the weekly Thursday seminar on quantum field theory techniques and particle physics. Two of my fellow students at this seminar were John Polkinghorne and Geoffrey Goldstone. Much later, in 1979, Polkinghorne left theoretical physics and trained for the Anglican priesthood. He has since become famous for his writings and lectures on the role of religion in scientific research, and how his personal beliefs in God and Christianity do not, in his opinion, conflict with the search for truth in scientific research. He was awarded the Templeton Prize in 2002 "for progress towards research or discoveries about spiritual realities." This prize, established by the British-American entrepreneur Sir John Templeton in 1972, is awarded annually with the stipulation that the cash award be slightly larger than the Nobel Prize, which is currently valued at almost $1.5 million. The Templeton Prize is awarded to people in many different fields and of different religious backgrounds, but the awards to scientists have been criticized as encouraging the undermining of rational scientific thought.

My other fellow research student at Trinity, Geoffrey Goldstone, showed early brilliance as a theoretical physicist. At the time of the annual Trinity fellowship competition, I discussed with him how he was faring with the preparation of his required research essay. He told me in a casual manner that he would write one soon. I asked him when the deadline for the essay competition was, and he calmly responded that it was in a few days. I felt deflated by this news, knowing that I could never have met such a deadline, or treated it with such aplomb. Goldstone later collaborated with the celebrated Hans Bethe on problems in nuclear theory. In 1961, he

published a fundamental paper introducing the idea of spontaneous symmetry breaking in particle physics.*

Another student at Hamilton's seminars with whom I became
friends was Walter Gilbert, who would later switch to molecular
biology, become a professor at Harvard and win the Nobel Prize
in 1980 for his work on recombinant DNA. While at Cambridge, he
was also an accomplished particle physicist and contributed important papers to the subject in the 1960s.

In this talented company, I entered the active field of particle
physics.

*

Invariably, at each Thursday seminar, Dirac would sit in the first
row, a slim, nondescript figure in his tweed suit. Early on in the talk,
he would look at the speaker in his rather distant way and ask, "Is
this a three-vector or a four-vector?" This referred to a vector field
in quantum field theory either being in three-dimensional space or
in the four-dimensional spacetime of special relativity theory. The
question really had no relevance to the talk, but Dirac always asked
it because, being the senior professor, he probably felt obliged to
ask a question. We students had decided that one week the answer
would be "three-vector," and the next week whoever was speaking
would reply "four-vector." Dirac seemed completely unaware of this
conspiracy, and after the question was answered, he would promptly

---

*In particle physics the vacuum is the state of lowest energy. This state can have a
symmetry, described by certain mathematical groups. The spontaneous breaking
of the vacuum state symmetry, Goldstone discovered, creates quantum spin zero
massless particles, which were later called the Goldstone bosons. The spontaneous
aspect of the symmetry breaking can be pictured as a ball balanced on top of a hill.
The balancing is precarious, and any slight perturbation triggers the ball to roll
down the hill. Once the ball chooses a direction in rolling down the hill, then it is
said that the symmetry of the balancing ball has been spontaneously broken.

fall asleep, with his chin sagging onto his chest and his mouth open.

Uncannily, however, very near the end of the talk, he would rouse himself, stare intently at the speaker, and, whatever the subject of the lecture, would ask his second question: "Can you fit the electron into your scheme?" Again, most often we considered this to be a totally irrelevant question because the talk would be about some technical subject such as dispersion relations in the scattering of elementary particles, in which the electron did not play a role at all. Again we had a conspiracy going, with three possible answers. One answer would be: "I'm not sure, Professor Dirac." The second one would be, "No, I don't think it can be fitted into this scheme." And the third one would be, of course: "Yes, you can fit it into this scheme." Dirac seemed unmoved by whichever answer we gave, and after some other desultory questions were asked, the seminar would end.

Dirac was particularly concerned about the electron because he did not keep up with the parade of new particles being discovered in accelerators at that time, and so could not ask a question about any of them. And of course the electron figured very prominently in his celebrated work on quantum mechanics, in his famous Dirac equation and in his early development of quantum field theory.

At the beginning of my second year of research, at one of the theory group seminars, Hamilton announced that the half-dozen or so students who had completed their first year of graduate work had to speak at each successive seminar on an original research topic of their choice. This was a daunting assignment, for finding an original research topic was a serious challenge so early in our careers.

When my turn came, I talked about quantum field theory and Hilbert space. This was a technically difficult corner of quantum field theory that had caught my interest. At my talk, everybody was surprised to see that Dirac did not fall asleep, and instead of asking his standard question about whether the field was a three- or four-vector, he asked me a specific question about the notation I was using. In fact, I was using a notation that was popular among American quantum physicists, in which a round parenthesis was used to open and close a symbol in the equations of a quantum field theory. I explained this to Dirac. At this time, it was also common practice to use Dirac's notation, which he had invented, called the "bra" and "ket" notation, which of course meant "bracket." The "bra" was in the form of a triangular bracket to the left, and the "ket" was a closing triangular bracket to the right. Amusingly, Dirac was very curious about my choice of notation, and why I had not used the famous notation he had invented. Unlike Schrödinger, who took me to task for using Einstein's method of deriving the unified field equations rather than his own, Dirac did not seem upset about my choice of notation; he merely wanted to know why I was using it. I replied that I was using the American notation because it was used in several of the papers I had studied that were pertinent to my talk.

Later in my talk, Dirac was very interested in a particular result I had discovered in relation to Haag's theorem in quantum field theory. This theorem concerns the technical details of Hilbert space in quantum field theory. Most surprisingly, Dirac focused on that topic, and did not ask his famous question about whether I could fit the electron into my scheme.

As it turned out, three or four of the students who had given talks at the Thursday seminars in Dirac's presence received letters from the Board of Research Studies informing them that they were being sent down from Cambridge—they were no longer research students at Cambridge! They had received no prior warning about this. Thus, although the circumstances of our graduate research and supervision at Cambridge seemed very lax—allowing me, for example, to neither attend courses nor take exams—it turned out that this was an illusion. Giving a talk as a second-year student on a piece of original research was very difficult because most students had not yet acquired the knowledge or the originality in physics to *have* their own ideas and to meet such a challenge. To be "guillotined" after such a talk was an ultimate and abrupt sentence.

Many years later, I attended a talk by Dirac at the annual meeting of the American Physical Society in New York, when he was a visiting professor at Yeshiva University there. He brought up the very issue that I had raised in my talk years earlier at the Cambridge Thursday seminar—Haag's theorem and non-separable Hilbert space—which he had shown such an interest in then. He now called this problem the "deadwood" of quantum electrodynamics (QED),* meaning that it was a part of quantum field theory that should not be retained in the theory, but was not easily removed from it.

As Dirac was leaving the auditorium after his talk, I stopped him, shook his hand and said, "I'm John Moffat. You may recall that I was a student in your theory group at Cambridge. At a Thursday seminar, I once talked about what you now call the deadwood of quantum electrodynamics." Dirac smiled and said in his cryptic way, "Yes, I do recall that. Very interesting."

I had some infrequent but memorable interactions with the great Dirac in my student days at Trinity. One afternoon I was walking up the stairs to the students' study room on the second floor of the Arts School and met Dirac on his way down. He stopped me and said, "Oh, Moffat. I want to ask you a question. Can you explain to me what this particle is that they call the K-meson?"

I explained to him in some detail what the recently discovered unstable, electrically neutral K-meson was and its place in the lineup

---

*Quantum electrodynamics is the theory that quantizes James Clerk Maxwell's classical electromagnetic field equations. It is a quantum theory of charged particles, such as the electron, and particles that convey the electromagnetic interaction between them, namely, photons.

of elementary particles. He smiled graciously, thanked me for the information and continued down the stairs. I stood still a moment before continuing up the stairs, pondering how a physicist of Dirac's stature could not have been following the growing and exciting literature on the new elementary particles being discovered in accelerators almost every month.

Another time, Dirac came out of his office as I was leaving the Arts School study room, and again he stopped me and said, "Oh, Moffat! You know, I have been studying Einstein's gravity theory, and I came upon this fascinating mathematical result." We went into the study room and he wrote on the blackboard a set of mathematical identities—equations that are always satisfied in an identical way—involving the Riemann curvature tensor and the conservation of energy in Einstein's field equations.

"Ah, Professor Dirac!" I exclaimed. "These are known as the Bianchi Identities, discovered by the Italian mathematician Luigi Bianchi in 1880."

Dirac smiled and said, "Oh, now, isn't that interesting!"

This incident revealed to me two very intriguing things about Dirac: first, it was not only the particle physics literature that he was not cognizant of, but the standard literature on relativity theory as well; and second, he was so ingenious that he had actually discovered the fundamental Bianchi identities by himself.

*

Towards the end of the class, Dirac wrote another equation on the blackboard, paused, and walked to the window, looking out at the grey spring morning with his hands behind his back. We students sat and waited. Finally he turned and said, "We will continue next week." Thus ended another memorable lecture by Dirac based on his famous book published in 1930, *Principles of Quantum Mechanics*.

We all stormed out of the Arts School lecture room. I nodded at a fellow student, James, as we entered Benet Street and walked together down King's Parade towards Trinity College and lunch. It had started to drizzle and I turned up the collar of my jacket, rearranged my black gown over it and said, "I heard a rumour that Heisenberg might be coming to Cambridge this week."

James was a second-year research student like myself at Trinity College, who had chosen quantum field theory as his Ph.D. subject. He had a pallid, scholarly look, and a slightly stooped posture even at the age of twenty-one. He sniffed, looked up at the sky and said, "Well, it looks like we're in for another wet spring. As for Heisenberg, yes, he is coming this week. He's giving a public lecture at the Cavendish Maxwell Lecture Theatre Thursday evening at seven."

"Can't miss that," I said. "I heard about his visit to New York a month ago when he presented his new ideas on a unified field theory of particle forces. Apparently that didn't go down well with Bohr and Pauli."

We passed the market and the Senate House and entered Trinity Street, making our way single file down the narrow pavement, as students jostled one another to get to Trinity and St. John's for lunch.

"I wonder if Heisenberg's activities with the German bomb research during the war caused any rift between him and Pauli and Bohr," I ruminated as we walked. "Those two weren't happy about his activities with the Nazis during the war. But I hope it wouldn't influence their behaviour towards him and the physics he's doing. Heisenberg is always working on some important new developments. I'm looking forward to his talk."

We reached the entrance of Trinity College, walked in past the porter's lodge and entered Great Court. "Did you see the *British Times* last week?" James asked. "They actually published Heisenberg's unified theory equation below the obituaries."

"Hmm," I said. "Is that where it belongs?"

*

Thursday evening, after my usual dinner of baked beans, chips and tea at the Lion's Restaurant, I made my way to the Cavendish Laboratory early, to get a seat for Heisenberg's public lecture. Already some of my black-gowned compatriots were seated in the front rows of the gloomy, austere Maxwell Theatre. I sat in the second row instead of the first in order not to make myself too conspicuous, but close enough to see the great man in action. About twenty-five minutes remained until the lecture, and already the hall was filling up. It was an audience consisting of professors and students from different faculties, the spouses of professors and their families. This was obviously going to be a popular lecture, I thought.

A side door opened and in walked Professors Neville Mott and Paul Dirac, and a slightly built man in a grey suit with greying blond hair, a smile and twinkling eyes. The trio came over and stood near us, talking physics. Our group leader, Jim Hamilton, joined them. Soon he led the visitor over to where we were sitting, and in his usual breezy autocratic manner, said, "These are the students in our group studying quantum mechanics and quantum field theory." We stood up and were introduced in turn to Professor Heisenberg, who leaned over and shook our hands one by one in a friendly way. "I just shook the hand of the famous Werner Heisenberg!" I thought, thrilled.

When the amphitheatre was filled, the doors were closed. Mott introduced Heisenberg, enumerating his successes and telling how he had helped to develop quantum mechanics and had won the Nobel Prize. Then Heisenberg walked to the podium and began lecturing. He spoke in an animated way, with a slight German accent, about his new unified theory, waving his hands enthusiastically for emphasis and striding back and forth across the stage in front of the podium. I thought that he had probably always been enthusiastic about physics and that his love of the subject must be the driving force that led him to make such great discoveries. Now he was in his fifties and had undergone difficult times during the war, being part of the Nazi war machine's effort to make an atomic bomb. He had also suffered serious privations at the end of the war when Germany was occupied by the Allies. It showed on his face. In spite of all this, he managed to inject a sense of humour into his talk, and smiled at the audience as he spoke. I noticed that he wore a gold tie pin in the shape of an h-bar, which stands for Planck's famous constant divided by two pi. This must have been a private joke, or perhaps a personal totem, for the h-bar was a significant element in Heisenberg's famous equation describing the uncertainty principle in quantum mechanics.

When the lecture ended, Mott invited questions from the audience. There were a couple of questions from lay people about developments in quantum mechanics, and how Heisenberg came to the idea of his unified field theory.

Then, suddenly, Dirac stood up and asked, "Werner, can you fit the electron into your scheme?" We students in the first and second rows looked at one another and couldn't help smiling at the familiar question, but the moment was disturbing too, because in Cambridge academic circles it was unheard of that a professor would seriously question another professor at a public lecture, particularly a speaker as renowned as Heisenberg. This was not cricket, not the English way of doing things. Dirac, widely referred to as "the silent physicist," normally never asked any questions at public lectures, and, in fact, was known for not expressing his views about anything at all. He preferred to hide out at his house on Cavendish Road, repairing his old Rolls-Royce and thinking about physics in his inimitable way.

Heisenberg stopped his pacing and suddenly looked quite pale in the face, staring uncomfortably at Dirac. We had heard from a professor who had visited Columbia University recently, where the meeting between Heisenberg, Pauli and Bohr had taken place, that this very point was the downfall of Heisenberg's unified theory as far as Bohr and Pauli were concerned. It was at this meeting that Niels Bohr had uttered his much-quoted comment: "Werner, your theory is crazy, but it is not crazy enough!" Pauli, too, had berated Heisenberg in his normal aggressive way, and the meeting had ended with Heisenberg depressed and upset. With the exception of the war years, when there was little contact between the two men, Heisenberg had often been subjected to Pauli's criticisms throughout his career, particularly when he was developing quantum mechanics in his early twenties. But Pauli had treated him particularly badly at the recent New York meeting because, in fact, he could *not* fit the electron into his scheme. And of what use is a unified theory without any electrons?

We all turned around to look at Dirac some rows back as we waited for an answer. Even Dirac, in his unworldly way, was beginning to look uncomfortable. He had simply asked Heisenberg whether the electron could fit into his scheme because that was his stock question. Knowing Dirac, he had probably not discussed Heisenberg's meeting with Bohr and Pauli with the professor who had been at Columbia and indeed was most likely unaware of the event.

The silence continued, and the audience waited. Finally, Heisenberg shook himself out of his reverie and said, "Well, Paul, this is the one serious problem I have with this theory. I cannot yet fit the electron into the scheme." The hushed silence of the audience was broken by murmurings. It was clear to everyone that Dirac had unwittingly hit the bull's eye with his question. It was also clear to Mott that Dirac had precipitated a socially disastrous situation.

There were no further questions, and the public lecture broke up. As we students left, we observed Dirac, Mott and Hamilton gathered around Heisenberg, soothing the wounds of our honoured guest.

As James and I walked into King's Parade past King's College, he laughed loudly and said, "Well, I wonder if Dirac is going to ask this question next Thursday at my talk."

I said, "Yes, indeed, that would be an interesting development, to see whether Dirac finally realizes that his question is irrelevant. But then again, it turned out that it was very relevant for Heisenberg!"

In contrast to Einstein's attempts to discover a unified field theory based on classical field equations for gravity and electromagnetism, Heisenberg's scheme only attempted to unify the subatomic forces, that is, the electromagnetic force, the strong nuclear force and the weak interactions, excluding gravity, and he used the techniques of relativistic quantum field theory. In my work on unified field theory, like Einstein, I initially left out the subatomic forces, although in Copenhagen I did attempt to make Einstein's unified field theory into a relativistic quantum field theory with the nuclear forces.

Heisenberg had put his group in Munich to work on his unified theory, and since he was the director of the Max Planck Institute there, everyone in the group was forced to work on this subject, whether they believed in the validity of the project or not. Unfortunately, the electron seemed to sit by itself in an antisocial way in this scheme because of its small mass, and it did not interact strongly with the other particles, such as the nucleons, the pi-meson, the K-meson and the rest of the strongly interacting particles discovered in the 1950s. This antisocial behaviour of the electron showed itself in Heisenberg's scheme. He had a method of calculating the masses of elementary particles from his equations, and the electron mass simply did not come out of the calculations. Heisenberg's scheme, which he claimed was a unified theory of particle physics, failed because clearly the electron, an essential stable elementary particle that played an important role in the theory of weak interactions, was simply not present in Heisenberg's model.

As it turned out in the following years, after much effort by those working with Heisenberg at the institute in Munich, the theory did not live up to his expectations. After his death in 1976, only a few devotees continued to try to solve the problems posed by Heisenberg's unified field theory. With younger physicists coming along and wanting to engage in more promising research, the project was eventually abandoned. Heisenberg's immense success with the discovery of the uncertainty principle, which is at the foundation of quantum mechanics, and his development of particle-based matrix mechanics, which was an alternative to Schrödinger's wave mechanics, had made him one of the most influential figures in postwar physics in Germany and internationally. But he was criticized by his peers both abroad and in Germany for demanding that his group, and anyone who was hired in Munich, or indeed at other physics institutes in Germany, should work on his unified theory project, which was not considered a worthwhile effort by physicists such as Bohr and Pauli and others in the United States.

In contrast, Dirac worked in isolation, rarely collaborating with other physicists, and had little influence on the progress of physics at Cambridge, despite the fact that he was just as famous as Heisenberg for his part in the development of quantum mechanics. Heisenberg won the Nobel Prize in physics in 1932, and the following year Dirac and Schrödinger shared the prize.

*

By 1957, near the end of my studies at Cambridge, I had to concern myself with my prospects of finding a job. I was then married to Bridget, who worked as a secretary at a firm outside Cambridge. We had hopes of starting a family someday, and I needed to find a secure position. There wasn't a great demand around the world for physicists specializing in gravitational theory. But I hoped that my recent work in particle physics would bolster my resumé. I applied for a government fellowship at the U.K. Department of Scientific and Industrial Research, and also for a position at the British Atomic Energy Laboratory at Aldermaston. Alarmingly, I had received notification that I could be conscripted into the army. The prospect of spending two years in the army, in the wake of the Korean War, did not please me. In my mind, this would be a serious impediment to a future academic career. I decided to ask Professor Dirac to write a letter of reference for the government fellowship I'd applied for and the position at Aldermaston. I would also ask him for a letter recommending that any army conscription be deferred. (As it turned out, military conscription continued in Britain until 1960.)

I made an appointment with Professor Dirac's secretary to see him at his house, a villa in one of the wealthier neighbourhoods of Cambridge. When I arrived, I spied a pair of legs in grey flannel trousers sticking out from underneath an old Rolls-Royce in the garage. I decided not to interfere with Dirac's mechanical repair work. I went to the door, rang the bell and was ushered in by a maid in a black-and-white uniform. She took me to the back of the house and invited me to wait in a sunlit drawing room overlooking a colourful garden. I saw no sign of Mrs. Dirac. I knew that she was the sister of Professor Eugene Wigner, a Nobel Prize–winning physicist at Princeton University. The story had gone around in the student circles at Cambridge that on occasion, when introducing his wife, Dirac would announce, "This is Wigner's sister."

I sat in an easy chair looking out at the garden, and listened to the birds singing. It was early spring, and sunlight streamed into the room through the French doors. Eventually Dirac appeared, with clean clothes and with washed hands. We shook hands cordially and he sat down opposite me. A silence ensued, during which I expected him to ask me why I had come to see him. But he said nothing. Eventually I said hesitantly, "Professor Dirac, I've come to ask whether you would write a letter of recommendation on my behalf. I have applied for a fellowship in the Department of Scientific and Industrial Research and a possible position at Aldermaston Laboratory doing nuclear physics research." I didn't add that the British government was busy trying to make an atomic bomb at Aldermaston.

Again, a silence ensued. We both stared out the French doors, and only the chirping of the birds broke the silence. Dirac simply smiled and said nothing. I moved uneasily in my chair and said, "Perhaps I should tell you about my current research activities. As you know, I gave a seminar when you were present on non-separable Hilbert space in quantum field theory." Dirac said nothing. "I'm planning to continue this research, and also further some work I have been doing on gravitation theory," I added.

I expected he would ask me some questions about my research, but he remained silent. "I also have to ask you for a letter so that I can get a deferment from military conscription. I've received notification from the military that I'm facing conscription after I finish my Ph.D."

The maid knocked on the door and arrived with a tray of tea and cakes. This was a welcome relief from the silent tension, broken only by my voice. We drank the tea in silence and I munched a piece of cake noisily. I began to feel that this interview was becoming ridiculous and surpassed any of my expectations regarding Dirac's reputation as being a kind of deaf-mute. He had talked to me in the past when he wanted information about some physics problem. Even then, the conversation on his part had been terse. Later in life, I would realize that Dirac showed symptoms of autism or Asperger's syndrome in his seeming inability to relate normally to people. Once when I was a visiting professor at the University of Texas at Austin, I attended a lecture given by Dirac when he was also a visiting professor there. After he finished his talk, Bryce

Dewitt, one of the senior professors in theoretical physics, known for his pioneering work on quantum gravity, stood up at the back of the room and asked a complex question that took at least five minutes to complete. "Do you agree with my impression that what I am asking is true, Paul?" he concluded.

Dirac stood in his tweed suit looking gaunt and much older than when I had known him at Cambridge. He contemplated what Dewitt had said, and then he merely said, "No."

Eventually the strain of the silence in the spring drawing room began to overwhelm me, and I stood up nervously and said, "Thank you, Professor Dirac, for seeing me. I hope that you can provide the needed letters of recommendation."

He accompanied me to the door, opened it and smiled at me. The maid came and nodded sympathetically. Without a word, I left. This was the strangest interview that I would ever experience in my life.

As it turned out, Dirac did provide a letter of recommendation for my government fellowship, and my application for a job at Aldermaston. And since I never heard another word from the army, I could only assume that he had been equally persuasive in a letter to them.

*

I would meet up with Dirac again in later years. In addition to the meeting in New York where he discussed the deadwood of quantum electrodynamics, he was often present at the Coral Gables conferences on particle physics held at Miami University, as he had moved to Florida State University at Tallahassee, after retiring from Cambridge. At one of these meetings, supersymmetry was the latest physics fad, and there was a special session on this new topic. Supersymmetry is a symmetry occurring in particle physics in which it is postulated that for every boson particle with integer spin, there is a corresponding fermion particle with half-integer spin. For example, the photon, which is a boson, would have a super-symmetric partner called the photino, a fermion. The theory was called supersymmetry because it would constitute the biggest sym-metry you can have in particle physics and spacetime. Supersym-metry was invented to get rid of certain technical problems in particle physics.

After hearing a talk on supersymmetry by an excited young physicist, Dirac stood up at question time and asked, "Why are we interested in this so-called supersymmetry, for no experiment has ever detected supersymmetric particles?" This was typical of Dirac's literal and logical approach to physics. In spite of his preaching that physics theories should be elegant and beautiful, he still maintained that such "beautiful" theories had to be verified by experiment.

One afternoon during that Miami conference, all the attendees congregated in a courtyard outside the conference auditorium for a group photograph. While the photographer was preparing his equipment, a journalist from a local newspaper interviewed Dirac. Behram Kurşunoğlu, a Turkish-born physicist who was director of a theoretical physics research institute at Miami University, organ-ized the annual Coral Gables conferences and idolized Dirac, was hovering over Dirac and the journalist in his usual protective man-ner. I was standing nearby and overheard the conversation.

Kurşunoğlu said to the journalist, "Professor Dirac is one of the greatest physicists of the twentieth century."

The journalist turned to Dirac and asked him, "Do you agree with this assessment, Professor Dirac?"

There was a long silence, and then Dirac said, "No."

Several seconds passed while Kurşunoğlu and the journalist di-gested Dirac's response. Then the journalist asked, "Professor Dirac, why would you not agree with this assessment of your career?"

Dirac brooded on this question and then answered, "Because I have only won one Nobel Prize." Dirac was referring to the fact that the American physicist John Bardeen had won two Nobel Prizes in physics—one in 1956 and the other in 1972—doing one better than Marie Curie who had won one prize in physics, but her second prize only in chemistry. In his literal way, Dirac found his single prize lacking.

Many years later, Dirac and his wife, Manci (Wigner's sister), attended a conference at Abdus Salam's International Center for Theoretical Physics in Trieste, Italy, where I was also an attendee. We all stayed at the Adriatico Palace Hotel on the waterfront, with the marina and its splendid yachts nearby. One evening before dinner, I was sitting in the hotel lobby withKurşunoğlu and Manci Dirac. This was the first time I had met Dirac's wife. She was a strong-looking, assertive woman who was not particularly attractive.

Eventually Dirac appeared out of the elevator in his perennial brown, crumpled, three-piece tweed suit. Manci frowned at him, and said sharply, "Paul, you are so stupid! You can't even put on your own trousers." It came out in the conversation that she had had to help him get dressed prior to her leaving the hotel room. Dirac looked sheepishly at us and smiled, and did not respond to this comment. I was surprised but not shocked by Manci's behaviour, because I had heard that she could be quite sharp and critical towards her famous autistic-savant husband. Kurşunoğlu looked upset and said, "Mrs. Dirac, how can you call one of the greatest physicists of the twentieth century stupid?"

Manci turned on Behram and barked, "He's my husband. If I want to call him stupid, that's my business." Behram had nothing further to say to this.

On the other hand, Manci could be very supportive of Dirac. After his retirement from the Lucasian chair at Cambridge, the university took away his parking space at the Cavendish Laboratory, which he had had for decades, and even worse, took away his office at the Arts School, forcing him to work at home. Manci wrote outraged letters on Dirac's behalf to the Cambridge administration. Although the letters certainly proved her support for her husband, they had no effect on the administrators' treatment of him. Neither did the fact that Paul Dirac was, arguably, the greatest British physicist since James Clerk Maxwell and Isaac Newton.

10

A B D U S   S A L A M

F RED HOYLE had left Cambridge in 1955 for the United
States to continue astrophysical research related to his work
on the composition of stars. Thus, at the end of my first year
at Trinity, I suddenly found myself without a supervisor. According
to the university rules, a research student at Cambridge had to have
a supervisor. As my research was going reasonably well, obtaining
a supervisor would be a mere formality because I would continue
working on my own.

The university proposed that I be supervised by Abdus Salam,
who was a young fellow of St. John's College and a newly appointed
lecturer in the university. Fortunately, Salam agreed to take me
on. At the time, he was not particularly interested in relativity and
gravity. He was busy researching quantum field theory and particle
physics and building up a particle physics group at Cambridge in
collaboration with his former supervisor, Paul Matthews. Consequently, my supervision under Abdus Salam turned out to be as
haphazard as it had been with Fred Hoyle. A session most often
consisted of my meeting Abdus at the Arts School on Benet Street
and striding at breakneck speed through King's Parade and Trinity
Street on the way to St. John's College, black gowns flapping, with

me breathlessly explaining my pursuits in relativity and gravitation to a silent Salam. We would part at the gates of St. John's College, and Salam would dart in to tutor undergraduates in his rooms.

He was known in Cambridge as Abdus Salam, but his real, unanglicized name was Ab-us-salam. Like his other students and physicists at the university, I called him Abdus. He had been born in the part of India that after partition became Pakistan, and his father was an Ahmadi Muslim, a sect that was subjected to much persecution in the Islamic world because its believers claimed that their late-nineteenth-century founder was the long-awaited Messiah, or Mahdi. Salam had married his first cousin and they could often be seen walking through Cambridge, she following him several yards behind, dressed in a black chador. I later learned that part of Salam's motivation for being a physicist was that as a deeply religious person, he believed that any success he achieved in theoretical physics would contribute to revealing the secrets of Allah.

This is not to suggest that Salam was interested in physics primarily as a means towards a higher goal. On one occasion when we walked rapidly through Cambridge together, he commented that it would be interesting to know how physics would look fifty years hence. This was a strong indication to me that he, like Pauli, was passionately interested in seeing how the fundamental theory of matter would develop and how he could contribute to a deeper understanding of particle physics.

It was understood from the beginning that my supervision by Salam would be on a purely formal basis, for with the exception of his collaboration with two student-colleagues, John Polkinghorne and later Walter Gilbert, Salam did not collaborate with students. Such a pro-forma supervision arrangement was not untypical at Cambridge, where students were often left at arm's length by their supervisors. It was a sink-or-swim situation, for without much guidance, you had to learn to be independent in your thinking and try to succeed on your own. I was quite happy with this arrangement.

I met Salam in the mid-1950s, before he became famous. He was a handsome man in his thirties then, with piercing dark brown eyes and a well-trimmed moustache. He usually dressed formally, in a three-piece suit and St. John's College tie. Salam had been a highly successful undergraduate at Cambridge, and had achieved the status of wrangler in the notorious tripos exams, which I had successfully avoided. The famous Cambridge tripos was in three parts, and it normally took four years to study for these exams. They originated in 1641, and consisted of candidates sitting on three-legged stools—hence the term "tripos"—with the examiner firing questions at them. They in turn "wrangled" or debated with the examiner as they answered the questions. In 1794, Cambridge University officially instituted the tripos exams as written exams. Those with the highest scores were called wranglers, and the one with the highest score was called the senior wrangler. Salam also did very well in the special physics tripos exam, and he carried out experiments using the old apparatus left over from the Rutherford era of the Cavendish Laboratory, which had been used to investigate the structure of the atom. Prior to his becoming a lecturer at Cambridge, Salam had spent a year at the Institute for Advanced Study in Princeton, where, in collaboration with his former supervisor, Paul Matthews, he had solved a difficult problem in quantum field theory called the overlapping divergence problem.

It was inspiring to see how passionate Salam was about physics. During my time at Cambridge, he gave a series of lectures on the strong force in particle physics and what is called dispersion relations. These relations refer in general physics to the index of refraction of a medium, such as water, being a function of the frequency of waves in the medium. I spent a lot of time in those days in the upper gallery at the Arts School library, poring over microfilms, turning a little wheel to read them. These microfilms were a pre-liminary copy of *The Theory of Quantized Fields* by the Russian physicists Nikolay Bogoliubov and Dmitry Shirkov, which was Salam's holy book on the subject. At one classroom session when Salam was lecturing and writing on the blackboard, he kept ignor-ing certain numerical factors, which made his calculations hard to follow—and I knew what those factors should have been, from reading the Bogoliubov-Shirkov book. I stood up halfway through Salam's lecture and complained to him about the lack of rigour in his calculations, turned on my heel and walked out. Salam never showed any anger at me for this arrogant behaviour. In fact, it seemed to me that I had won more respect from him as a physicist.

*

During one of my early sessions with Salam, I told him how I had corresponded with Einstein prior to my arriving at Cambridge. His eyes lit up. "You corresponded with Einstein?" he said. "We have to do something about this."

"What do you mean?" I asked.

Salam's eyes narrowed. "You have to get these letters copied and send them out to the major universities and physicists like Dyson, Oppenheimer and Wheeler in America," he said.

I was astonished at this proposal, alarmed at the idea of adver-tising myself in this way. Salam noted my discomfort and said, "Moffat, you have to learn to sell yourself. The physics world is a competitive place."

Little did I know that this statement portended events in Salam's own life, when he would be fiercely competitive and lobby research groups and even the Nobel committee, promoting his ideas. I and other graduate students at Cambridge idolized Paul Dirac, who seemed able to keep himself completely aloof from academic poli-tics. He published his brilliant ideas and achieved fame without getting his hands dirty in the fiercely competitive world of physics. I realized that this was probably a naive and overly idealistic attitude to adopt, especially in view of my lowly position at Cambridge, but I could not make myself follow Salam's advice.

About a week later, as I was descending the stairs at the Arts School, Salam was ascending them breathlessly. He stopped me and asked, "Well, Moffat, have you sent out those Einstein letters to the people I suggested?"

For a moment I said nothing, and then hesitantly said, "No, I haven't done anything about it."

"Why not?" Salam asked.

"Because I feel uncomfortable about it."

"How do you expect to make a name for yourself in physics with this kind of attitude?"

I gazed at Salam and remained silent. After a few moments of frowning at me, he climbed up the stairs impatiently.

*

Salam was in a unique situation at Cambridge, being the only notable theoretical physicist from the Indian subcontinent. He was, in fact, the first Pakistani-appointed professor in the British university system after Pakistan was formed in 1947—and he undoubtedly at times felt discriminated against by the English academic establishment. Another wrangler at Cambridge, John Meggs, was working on a rather original approach to quantum field theory. He was a tall, gangly, bespectacled Englishman. The rumour among us research students was that he and Salam had had a serious difference of opinion about Meggs's research, and Meggs had in anger uttered a racist epithet to Salam, which had greatly upset him.

Meggs eventually joined the army and set aside his final Ph.D. defence for a year. One day I happened to be present at the Arts School when Paul Matthews and Salam were standing outside a room waiting to examine John Meggs for his Ph.D. Richard Eden, a former student of Dirac's, appeared and there ensued a heated discussion between the three of them, which I witnessed. Eden was warning Salam and Matthews that there would be serious repercussions if they did not award Meggs his Ph.D.; Meggs was an outstanding student at Cambridge, he said, and it would cause a scandal if they failed him. The undercurrent that I detected in Eden's comments was that Salam should not allow Meggs's possible racism to influence his judgment of the Ph.D. candidate's intellectual achievements. Salam and Matthews did award Meggs his Ph.D., and Meggs subsequently left Cambridge for a position in industry.

*

In 1957, Salam was appointed professor of applied mathematics at Imperial College London. Again, I was left without a supervisor, and this was at the awkward time when I was almost finished with my thesis. The university suggested that a junior fellow at Cambridge, David Candlin, serve as my supervisor for the remainder of the year, the same David Candlin who had been one of the victims of Pauli's grilling during his visit to Cambridge, and who was just two years ahead of me, having only recently earned his doctorate. Just as with my previous two supervisors, Candlin and I had no serious discussions about my research. He was engaged in research in quantum field theory, a topic that I also became involved in, working on quantum field theory along with my Ph.D. work on modifications of gravitation theory and the motion of particles in Einstein's theory. The interlude with Candlin was brief, and not very satisfactory, because we had so little contact with one another. Without the respect and fame of a Hoyle or Salam backing him up, Candlin seemed nervous about my unusual status at Cambridge—a Ph.D. student without a prior degree.

Although I still had not earned my degree, I had to make plans for my future after Cambridge. The letters that Dirac must have written on my behalf began to bear fruit. One day a portly gentleman in a black suit and bowler hat, carrying an umbrella, arrived at Bridget's and my small flat in Cambridge and said that he was with the security service at Aldermaston Laboratory. He interviewed me intensively about my past, no doubt to ferret out whether I had plans of becoming a spy and stealing secret atomic bomb information from Aldermaston. Needless to say, I had no such plans, and after scribbling in a notebook, he seemed satisfied and left. To me, the job at Aldermaston was definitely second choice. I earnestly hoped to receive a fellowship from the government in order to continue physics research after earning my Ph.D.

Not long after this, I took the train into London for an interview at the Department of Scientific and Industrial Research (DSIR), which awarded those research fellowships. I was discouraged to find at least ten other young hopefuls waiting on benches in a long, narrow corridor, all having applied for the two available fellowships. I felt very tense. The door opened and we were ushered into a large room where professors were seated at individual tables with little plaques announcing their titles and affiliations. I was seated in front of a tall, grey-haired, bespectacled gentleman wearing a light brown suit and a college tie from some university in England that I did not recognize. He began interviewing me about my academic background and raised his eyebrows when he heard that I didn't have an undergraduate degree. I laid out reprints of my papers published in the Cambridge Philosophical Society's journal on the table between us. He skimmed through them and seemed impressed. I explained that I expected to finish my Ph.D. shortly.

I left the building and walked down Oxford Street towards Piccadilly Circus feeling despondent, not expecting to win this fellowship with such competition from the other applicants, who, I imagined, possessed more conventional academic backgrounds. I took the train back to Cambridge and the flat Bridget and I were renting across the street from a neglected little cemetery. We had just one room with a pullout bed, and a combination kitchen and bathroom, with the bathtub placed against a wall next to the little gas stove. There, I continued typing my thesis on an old Royal typewriter, using carbon paper to make a copy. I laboriously printed in the equations with a pen between gaps I had left in the text. In those days long before word processors, writing a thesis was an agonizingly difficult task. In between bouts of typing, I would sit and stare out the window at the grey headstones across the street, attempting to feel positive about my future.

I was in close touch with my parents during this time. My father had recovered from tuberculosis and worked as a draftsman at an aircraft company outside Cambridge. On his way home in the late afternoons, he would often stop by to visit. He drove an ancient black sedan that constantly needed expensive repairs. My parents rented out a room in their house on Oxford Road, and my mother prepared meals for the lodger, who was usually a student from abroad.

*

Near the end of 1957, I submitted my Ph.D. thesis to the university and appeared before my examiners, the relativists William Bonnor, who was by this time a professor at the University of London, and William McCrea. Remarkably, these were two of the professors who had interviewed me when I had arrived in England three years before. After the intense verbal exam was over, I learned from McCrea, who met with me privately in one of the rooms at St. John's College, that they weren't happy with the way I had written the thesis, and wanted me to do more work on it. This dampened my spirits considerably, as more tedious work lay ahead. But not long afterwards I received a letter from the Department of Scientific and Industrial Research informing me that I had won the fellowship I had wanted so badly! This gave me renewed motivation to get back to work and make the required changes on the thesis. And when the job offer came from Aldermaston, I was relieved to be able to turn it down.

I later learned that Roy Kerr, who was finishing his Ph.D. at about the same time, also had Bonnor and McCrea as examiners, and he suffered the same fate as I did. I knew that Roy was not fussy about English grammar or the presentation of his work, and indeed, McCrea demanded that he not only rewrite parts of his thesis, but even put in literary quotations at the beginning of each chapter.

Abdus Salam was now a full professor at Imperial College London, where he had been asked to form a new group concentrating on particle physics and quantum field theory. Even though I had not yet finished my Ph.D., he invited me to join him there with my newly won government fellowship and become his first postdoctoral fellow. I went to London alone at first, without Bridget, who continued in her job in Cambridge, and I moved into a flat near Kew Gardens. The ensuing few months were not a happy time on my own, as I laboured away every day, rewriting parts of my thesis. I worked alone, without advice from either Bonnor or McCrea, and Salam was too busy forming his new group to be of any help. My parents seemed disappointed when they heard the news about my thesis. At the Cambridge bus station on the day I left for London, my father reprimanded me for not being more successful. Within a couple of months Bridget left her job in Cambridge and moved to be with me at the flat in Kew.

In the thesis that I originally submitted, I had developed a modification of Einstein's gravity theory, general relativity, based on a different geometry than Einstein's, a complex symmetric Riemannian geometry. McCrea and Bonnor had not thought this was enough to constitute a Ph.D. thesis, so in my revised version I included all my work on the motion of particles in Einstein's gravity theory. This meant that I had to include the problems I had discovered in the paper by Einstein and Infeld published in the *Canadian Journal of Mathematics* in 1949, and my attempts to resolve these problems in a positive way.

When I was finally finished rewriting the thesis, I arranged with McCrea to have my second Ph.D. exam at Cambridge. I took the train back to Cambridge, stayed with my parents for a few days and then met my examiners on the prescribed day in rooms at St. John's College. The exam lasted three hours, with Bonnor and McCrea asking one question after another. This time, they were very concerned about the issue of my criticism of the Einstein-Infeld paper. In fact, they were rather disconcerted about how to handle it. I had provided detailed calculations in the thesis, showing how the problems had arisen in the paper and also how to solve them. My examiners couldn't find anything wrong with my calculations, but they felt nervous about the political implications. Einstein was dead, but Infeld was in Warsaw—I had met him at the Einstein Fest in Bern before I had encountered the mistake in the famous paper—and he had never responded to my subsequent letters. Moreover, my modification of Einstein gravity was for that time quite a radical piece of work, which rang alarm bells for physicists as conservative as Bonnor and McCrea.

After three exhausting hours, McCrea turned to Bonnor and said, "All right, this is enough." They sent me out of the room and I sat outside waiting nervously, fearing that my postdoctoral fellowship, my father's anticipated approval and my entire future were about to go down the drain. Finally, they called me back into the room, shook my hand in turn and announced that I had passed the examination.

The first Ph.D. exam, the rewriting of the thesis and the gruelling final exam had been the most stressful series of events I had experienced in my short academic life. I promptly lost my voice and was unable to speak for several days. I also developed a nervous-stomach disorder during those months, which has stayed with me ever since, a frequent and painful reminder of my initiation into physics.

On the bright side of things, however, I now had a Ph.D. from Trinity College, Cambridge. Being the first student in the history of Trinity College to be awarded a Ph.D. in theoretical physics without an undergraduate degree marked quite an achievement. I now officially joined Imperial College London, and, fortified with my doctorate and feeling more secure about my future, I buried myself in my research.

Bridget and I moved from the Kew flat into one looking out onto Kensington Gardens, not far from the Albert Memorial and concert hall, and only a short walk from Imperial College in South Kensington. We were living on my meagre government fellowship and her salary as a secretary in a bank in South Kensington, and we felt lucky to have secured this small but pleasant one-room flat. Early in the morning, we would be awakened by the Royal Guards in their magisterial uniforms with glittering helmets and white feather plumes. Mounted on their magnificent horses, they clattered down the street on the way to Buckingham Palace.

11

IMPERIAL COLLEGE
LONDON

BDUS SALAM'S new group was initially housed in the
Mathematics Department at Imperial College London. This
was in European-style old buildings, which had originally
been the home of the Royal College of Science and the Royal College of Art. To prepare for Salam's arrival, the university had attempted to transform the shabby quarters of the Mathematics
Department into a grander suite more appropriate for a distinguished young professor with a promising new research group.
Salam's office was luxurious, with a large Persian carpet on the floor,
a mahogany desk at the far end and a blackboard on one wall. Parti-
tions had been set up in other rooms to section off smaller offices.
My office was in one of these, adjacent to Salam's, and had been
newly furnished with a Danish modern teak desk and an attractive
easy chair. Although there was no Persian rug on the floor, it was
certainly the most splendid working space I had so far enjoyed.

All this attention paid to Salam by the Imperial College administration caused some consternation among the older mathematics
faculty, who still inhabited ancient offices with dilapidated furniture.

A week after I began work in my new office, I came into the building one morning to find half a dozen of these older professors gathered at my door, peering in at my furniture and muttering derisively. The next day when I opened my door, I found that my beautiful furniture had been removed, and a battered old desk and scuffed chair stood in their place. When I showed Salam this evidence, he went into a rage, considering it a personal insult. He assured me that he would take care of the problem. Some days later I opened the door to my office and discovered with amusement that my smart furniture had been returned. Such was the clout of the new applied-mathematics professor at Imperial College!

Indeed, Salam in his impressive office was like the sun in a solar system of satellite planets. After my appointment as post-doctoral fellow, John Clayton (J.C.) Taylor, also from Cambridge, whose Ph.D. thesis had been on fundamental problems in quantum field theory, joined our group as a lecturer at Imperial College. Ray Streeter, who had performed brilliantly as an undergraduate at Imperial College, was one of Salam's new graduate students. After some months, Peter Higgs was appointed as a second post-doctoral fellow, and we shared my well-appointed office. Peter would go on to become famous for inventing the Higgs particle. Eventually, in 2008, a $9 billion proton collider, the Large Hadron Collider (LHC), would be completed at CERN, Geneva, one of its main purposes being to search for Peter's particle, which almost everyone believed to be the linchpin of the standard model of particle physics. Peter was a quiet, diffident, pleasant person; oddly, we didn't discuss much physics together.

All of us would orbit Salam's grand office, attempting to meet with him to plan future research projects. We expected Salam to provide us with research guidance, but as it turned out, our sun had only one stock-in-trade problem, which he handed out to graduate students and post-doctoral fellows. This consisted of following a suggestion of Julian Schwinger, a Nobel Prize winner at Harvard University, who had formulated a way of engendering the masses of elementary particles. The problem of what produces the masses of elementary particles goes back to the early days in the development of electromagnetism. The Dutch physicist Hendrik Lorentz proposed that the mass, or energy, of an electron is purely of electromagnetic origin from Maxwell's field equations. Schwinger proposed that instead of the energy of electromagnetic fields being the origin of mass, a new field that he proposed served this purpose. At that time, in the late 1950s, the subject was just developing. I studied the Schwinger paper, and after a few days decided that it wasn't my cup of tea. However, Ian Gatland, a new graduate student, seized upon this problem and made it the topic of his Ph.D. thesis. And Peter Higgs, of course, later pursued the idea of generating masses by extending Schwinger's ideas, and predicted the existence of what would be named the Higgs particle. Today, the standard model of particle physics proposes that the field associated with the putative Higgs particle is responsible for producing the masses of all the elementary particles.

We students and post-docs learned that Salam regularly attended an Ahmadi mosque in London, and he had accommodations in Putney, not far from the college, to which we were never invited. Whereas at Cambridge, Salam's wife had been frequently seen walking with her husband in the evenings, in London we never saw her. She and Salam spent considerable time apart; while he pursued his academic career, she often spent time in Pakistan with their two daughters.

Salam was very busy outside the department. He undertook a feverish schedule of travelling to and fro across Europe and the United States, giving lectures on his latest particle physics theories and promoting his new group at Imperial College. He was trying to make our group a centrepiece for particle physics research in the world physics community. For the rest of his career, this frenetic travelling would be part of Salam's lifestyle. Somehow he managed to give his lectures in between catching planes from London to distant lands. I wondered already at that time, when he was only in his thirties, how long Salam could continue this extraordinarily demanding schedule. It seemed there was little to offset his many hours of sitting in planes and airports. Although he was not over-weight, he was not interested in physical exercise.

While busy building up a reputation for his particle physics group, Salam continued his collaboration with Paul Matthews. Matthews had been born in India of missionary parents, and like other English physicists of similar background, such as Tom Kibble, who became a professor at Imperial College London, his personal history with the Indian subcontinent may have made him gravitate towards Salam, resulting in a lifelong friendship.

Almost every day at lunchtime, Matthews, now a professor in the Mathematics Department, John Taylor and I would meet at the faculty restaurant at Imperial College. At some point during the lunch, Salam would appear, and while eating a sandwich, would excitedly tell us about his latest physics idea. He had new ideas daily, and would use the three of us as sounding boards. After listening carefully to the daily idea, the Matthews-Taylor-Moffat trinity would proceed to demolish it. We almost always succeeded. But Salam, undeterred, always appeared the next day with another idea. However, now and then one of Salam's ideas would survive the lunch, and he would then dash off and pursue it further with great intensity. It would eventually surface in a physics journal as a joint paper with Matthews, who would have worked out many of the technical details. This partnership—with Salam usually providing the motivating idea and Matthews working out all the tedious calculations with the correct numerical factors—was an ideal collaboration.

At this time, in the late 1950s, Salam was upset about having missed the recent Nobel Prize for proposing the left-right parity symmetry violation in the weak interactions of particles, and he would frequently complain about it. This was a hypothesized breaking of the left-right spatial symmetry in the weak interactions of elementary particles responsible for radioactive decay. It was believed that when particles interact, they do so in either a right-handed or left-handed fashion. A parity operation would replace particles with their mirror images, the right-handed particles changing to left-handed ones and the other way around. In the electromagnetic and strong forces, parity is conserved: left-handed and right-handed particles behave the same under the influence of these forces.

But as Salam, among others, discovered, the weak nuclear force is asymmetric for right-handed and left-handed particles and thus violates parity: the interactions of the mirror-image particles behave differently than the original ones. In physics, this situation is represented by changing the sign of the x, y and z coordinates in three-dimensional space, which is the so-called parity transformation. Salam had come upon this idea and worked out some of its consequences for radioactive decay. In 1956, the experimental physicist Chien-Shiung Wu of Columbia University had observed parity violation in her laboratory, triggering the Nobel Prize of 1957, which was awarded neither to Salam nor herself.

How had this come about? Salam idolized Wolfgang Pauli, who was considered the chief justice in the particle physics community, apt to condemn any proposals that he considered unworthy. Salam had made the mistake of sending Pauli a letter with his new manuscript proposing that the left-right parity symmetry in particle physics was violated by the weak interactions, or beta decay. Pauli replied in his convincing way that left-right symmetry would never be violated in nature, and anyone who considered this possible was uttering nonsense.

However, two Chinese-American physicists, T.D. Lee and C.N. Yang, had already published a paper in 1956 predicting the left-right parity violation in weak interactions, and had convinced Wu to do the experiment. Pauli believed that this published paper was incorrect, and persuasively steered Salam away from pursuing the idea further. Lee and Yang were awarded the Nobel Prize for their work on parity violation in 1957. The Nobel Committee was criticized for sexism in overlooking Wu for the prize. Their passing over not only Salam, but a large handful of other physicists who had actually published on the subject, including Murray Gell-Mann and Richard Feynman, was more understandable.

At our Imperial College lunches, Salam would occasionally display his frustration at having missed this Nobel Prize. The fact that he had not published his ideas about parity violation set a rule in his mind for the future. Henceforth he would always publish any idea that seemed reasonably worthy of attention, regardless of whether he was sure it was right. He demanded the same of us in our research, chiding us always to publish our ideas, on the off chance that they would turn out to be important. One consequence of Salam's new rule was that he often published ideas in physics that did not measure up to his best work.

*

Despite Pauli's bad advice on the subject of parity violation in weak interactions, Salam continued to idolize him. Salam invited Pauli to Imperial College to visit his new group in 1959, and the students and post-docs were told to dress for the occasion. The day of Pauli's arrival, Salam entered my office looking his most dapper self, in what appeared to be a new, dark three-piece suit, white shirt and St. John's College tie, with his black hair and moustache gleaming. He informed me that I should attend to Pauli during his visit, to make sure that he was comfortable and had everything he needed.

This was the second time in a few years that I had been made a lieutenant in the ranks of those hosting the great visiting Austrian physicist, and I wondered what was in store for me this time.

The seminar room where Pauli gave his lecture was packed not only with students, physicists and mathematicians from our department, but also with curious people from other departments in the university. Pauli's lecture was peppered with his typical negative comments about other physicists and their theories. Afterwards, Salam asked me to go to his office and provide company for Pauli because Salam had to give a lecture. When I arrived at Salam's office, I stood leaning against the wall waiting, while Pauli paced back and forth across the opulent Persian rug, his hands clasped behind his back, muttering to himself and displaying his usual nervous tic, his bobbing head. I stood silently viewing this activity for perhaps as long as twenty minutes. All of a sudden, Pauli stopped in his tracks, whirled around, stared at me and shouted, "What are you doing here?"

I jumped in alarm and managed to stammer, "W-well. Professor Pauli, I hope I'm not disturbing your thoughts. I, er ..."

Pauli looked at me closely and then thundered, "Are you not the student who took me to lunch at Cambridge and answered back when I commented on your talk?"

"Yes, I'm Moffat."

"What are you doing here?" he asked, perhaps supposing I had wandered in off the street.

"I'm one of Salam's post-doctoral fellows."

"And what are you working on in your physics research now?"

"I'm working on different problems in particle physics." He viewed me suspiciously, before continuing his pacing of the office, and again becoming lost in his own unique physics world.

*

I had to start planning for my future again, as my two-year post-doctoral fellowship would soon be coming to an end. I was faced with the standard prospect of someone in my position of applying for a lectureship at some British university. I knew that if I took such a position, the significant teaching loads would cut severely into my time for research. One day, Paul Matthews approached me and asked how my research was coming along and what I proposed to do with my future. I said I was studying complex variable theory and its application to dispersion relations in particle physics. He looked at me critically and said, "Well, you'd better start producing some physics, Moffat. You haven't produced any papers since you arrived at Imperial College. Do you think this is good enough?"

I felt disconsolate at his criticism, and didn't know what to say. "Well, I do have some ideas that may eventually produce a paper," I said hopefully.

Matthews frowned and said, "Well, let's hope so, for your sake."

The challenges in getting my Ph.D. had affected my research productivity. Moreover, I was also in the process of learning a lot about particle physics. I didn't feel I was quite ready to produce original, useful work for publication.

I wrote to John Wheeler at Princeton and asked him whether he had an opening for a post-doctoral fellowship, or whether there were any other research positions in the States that he knew of that I could apply for. I had met Wheeler a year previously while he was visiting King's College London, giving a series of lectures on Einstein's gravitational theory. He had seemed very cordial and interested in my research on gravity. Wheeler responded kindly to my letter, saying that a new research institute was being formed in Baltimore, Maryland, called the Research Institute for Advanced Study (RIAS). He suggested I send a letter to the director, Welcome Bender, and ask whether there were any positions available. I did so, and not long after, I received a promising letter from Bender, asking me for details about my research. After some negotiations, I was offered a position as a senior research physicist at the institute.

Going to America would be a big new step in our lives. It meant leaving my parents and Bridget's mother, who were all living in England. However, the prospect of taking up this position with a reasonably good salary, without any teaching duties, was extremely tempting. I went to my critic, Matthews, and told him about this offer. He said, "What about the DSIR fellowship? Are you going to leave it just after one year? Besides, what is this institute anyway? Perhaps you'd be better off getting a lectureship at someplace like Birmingham or Manchester and pursuing your research in a proper English, established academic environment."

I found Matthews an intimidating person. His brusque behaviour stoked my sense of insecurity, which was already quite well developed, as I always felt as if I was on the outside of the academic establishment at Cambridge and Imperial College looking in. However, despite Matthews's misgivings and advice, I accepted the offer from Welcome Bender at RIAS.

But now I had to find the money to get us to the States. RIAS promised to reimburse me for my travel expenses when I arrived in Baltimore, but not to cable me the money ahead of time. My parents and Bridget's mother had no funds to help us. So I walked into a local Barclays Bank on Kensington High Street, around the corner from our flat—my home bank still being located in Cambridge—and made an appointment with the manager. The bank offices reminded me of clerical offices in a Dickens novel, with silent clerks sitting at dusty desks on high wooden stools, pens scratching. In my most appealing and persuasive manner, I explained to the bank manager in his gloomy office that I needed 60 pounds to buy one-way tickets on a ship to get my wife and myself to New York, as I was accepting a job as a physics researcher in Baltimore. Could the bank possibly lend me the money? I swore that I would pay it back as soon as I got to Baltimore and was reimbursed for the travel expenses.

The manager turned out to be a kindly gentleman. He told me that he had thought of pursuing a career as a mathematician at Bedford College, London, but gave up the idea and turned to the banking business instead, with much regret ever since. No doubt wishing to spare me a similar life of regret, he said with a smile, "I will see to it that the bank gives you this loan, and I wish you the best of luck in your future career."

Hardly believing my good luck, I shook hands with him and stepped out onto Kensington High Street. The busy traffic and spring sunshine reinforced my feelings of relief and new-found hopes for a future in America. Soon after my arrival at RIAS, I dutifully paid back the loan from the bank.

12

_____

BALTIMORE

I N THE SUMMER of 1959, Bridget and I sailed across the
Atlantic on an American liner and then took the train from
New York to Baltimore. There, we experienced for the first
time the semitropical heat and humidity that was typical of a
Maryland summer. I would soon have to shed my tweed jacket
and trousers and heavy shoes. As I had just turned twenty-seven,
I escaped being conscripted into the armed forces as a newly landed
immigrant in the States. Even though the Korean War had ended in
1953, military conscription was still being enforced, and non-U.S.
male citizens from ages eighteen to twenty-five who set foot on
American soil were not protected from it. Paul Dirac's influence
would not have helped me to avoid the American draft.

The Research Institute for Advanced Study, which everyone
called by the acronym RIAS, was situated in a large estate in north
Baltimore next to the one where the writer F. Scott Fitzgerald had
often stayed during the summer months with his wife, Zelda. Two
adjoining buildings on the property constituted the institute. One
housed the physics and chemistry research facilities and the other
the mathematics group, run by Solomon Lefschetz, a well-known
mathematician who had been a professor at Princeton University.

He was an energetic elderly man with two prosthetic hands, on which he wore black leather gloves. He had lost both hands and forearms as a young man in a laboratory explosion, when he worked at the Westinghouse company in Pittsburgh.

RIAS was a prototype of a non-profit research institute devoted to pure, fundamental research, yet funded by industry and/or the military. RIAS was funded by an aircraft company, the Glenn L. Martin Company (later Martin Marietta), which had its headquarters outside Baltimore.* U.S. military contracts supported much of the fundamental research at RIAS, as they did elsewhere in the late 1950s and 1960s in the United States. These were the post-Sputnik cold war days, in which there was a frantic effort in the United States to produce nuclear and other weapons to counter a possible attack by the Soviet Union. I applied for and received a generous research contract from the U.S. Air Force for my particle physics projects at RIAS. I continued to concentrate on particle physics research, rather than gravitation and relativity theory.

Bridget and I rented an apartment on the second floor of a house on Fifty-sixth Street in Baltimore, which at the time was the dividing line between the city's mostly black population and its white population. Being from England, we did not immediately realize that we had taken up residence on the front lines of the turbulent civil rights movement in the southern states. It wasn't long before we experienced first-hand what that strife was all about. One evening we went to a restaurant in a nearby mall for dinner, and were surprised to find that we were the only customers. Soon, an elegantly dressed black couple entered and sat at a table near us. The restaurant owner walked over to them and demanded that

---

*In contrast, the Institute for Advanced Study at Princeton, which was founded in 1930, was funded not by industry, but by the Bamberger family, and its first director, Abraham Flexner, hired prominent scientists such as Albert Einstein, Kurt Gödel and John von Neumann.

they leave. They refused. At that moment, a large demonstration of black people formed outside the restaurant, with banners proclaiming equality for Negroes, as the term was at that time. The crowd didn't voice their feelings loudly, but just stood outside the restaurant looking in. Bridget and I felt upset about this. We couldn't continue eating, and left.

We had planned to go to a movie at the mall after the abortive dinner. We bought tickets and went into the theatre, and discovered that we were the only people there. Again, three or four elegantly dressed black people came and sat near us. The same scenario as we'd witnessed in the restaurant was repeated. The theatre manager requested that they leave. They refused, and I guessed that another demonstration was forming outside. Nonetheless, the lights went down and the film commenced. When we left the movie house, we had to make our way through a large demonstration of African-Americans waving picket signs.

Some weeks after I joined the institute, a young black researcher from California who was conducting experiments in biochemistry arrived with his wife and two children. One afternoon there was a loud noise outside. I went to find out what had caused the disturbance, and saw that the young researcher's car was on fire. Not long after this incident, he and his family returned to California.

*

At the time, there were only half a dozen physicists at RIAS, including myself, doing research in theoretical physics. The head of the group was Louis Witten, the father of the future well-known physicist Edward Witten. At times, Witten junior would turn up, a toddler with a large head and pale complexion, with his mother. Among the other members of the theoretical physics group were Otto Bergman, an Austrian who worked on general relativity; an American, Philip Schwed, who worked on quantum field theory;

and Ian Gatland, a former student at Imperial College London, who had applied for a position at the institute after finishing his Ph.D. with Abdus Salam. I was also friendly with the metallurgist Henry Otte.

Philip Schwed was the senior physicist with whom I most often discussed my research. A brilliant theorist, he was a former Ph.D. student of Boris Podolsky, who had achieved fame by collaborating with Einstein and Nathan Rosen on the famous Einstein-Podolsky-Rosen paradox paper. Very sadly, Philip committed suicide during my time at RIAS. I never discovered what had driven him to it.

Ian Gatland and I began collaborating on particle physics, working directly with the data coming out of high-energy accelerators. In particular, we worked on the topical issue of Regge poles, named after the Italian physicist Tullio Regge, which was a clever way of calculating the scattering amplitudes resulting from the collisions of elementary particles in accelerators.

Every week I would attend the seminar on particle physics and quantum field theory held in the Department of Physics at Johns Hopkins University at the campus in downtown Baltimore, where I became friends with the physicists Gordon Feldman and Robert Fulton.

John Ward, a well-known physicist from Oxford University, had been appointed professor at Johns Hopkins, and in his peripatetic career in physics, would only remain there for a year before moving on to Australia to take up a professorship at Macquarie University in Sydney. John Ward had achieved fame through his brilliant work on quantum field theory and renormalization theory.[*] I was intrigued by his work, and upon being introduced to Ward at the weekly seminar in the Physics Department at Johns Hopkins, I

---

[*] A mathematical identity in renormalization theory, related to the requirement of gauge invariance in quantum electrodynamics, was known in the literature as the Ward identity.

invited him home for dinner. We set a date for a week later, and Bridget spent some time preparing a dinner with wine.

Remarkably, Bridget and I spent the entire evening listening to Ward talking about himself in a seemingly endless monologue. He never once addressed my wife, and, in fact, barely looked at her as we all sat in the living room, which was hot and humid since we did not have an air conditioner. Ward related how he hated Oxford, and had once considered obtaining some explosives and blowing up the university. I realized that Ward showed severe misogynist tendencies and was possibly a misanthrope; he seemed quite mentally unbalanced.

Some months after our strange dinner, I attended the spring meeting of the American Physical Society in Washington, D.C. In those days, these were large gatherings. I was standing in the lobby of the hotel where the conference was being held, at the edge of a crowd of physicists the first morning of the meeting, when John Ward walked in and saw me. He immediately came over and started talking about himself and how he was going to give a talk about the great discovery he had made with Salam that solved the problem of unifying the strong, weak and electromagnetic forces. Suddenly, to my amazement, he started making jerky movements with his arms. He darted away to the far side of the hotel lobby without any explanation.

The news about this great discovery that Ward and Salam had made, and that Ward was presenting their results at a major talk in one of the main ballrooms, spread throughout the meeting. When the time came for the talk, the ballroom was jammed with physicists.

We all waited for Ward to arrive to give his talk, the first in that session. He finally turned up, carrying a sheaf of papers, and hesitantly approached the podium. He stood looking out at the large audience for several minutes while there was a hushed silence in the ballroom. Finally, Ward mumbled a few words about how this discovery that he had made with Salam was going to change physics dramatically—they had finally discovered the unified theory of the forces of nature except for gravitation, he claimed. He then abruptly burst into tears. The whole audience witnessed this brilliant physicist having a nervous breakdown in front of us. Sidney Coleman, a young Harvard faculty member who had been a student of Murray Gell-Mann, rose from his seat in the first row. He approached the podium, talked with Ward, and he and another physicist led Ward out of the ballroom. I felt a sense of shock and sadness that John Ward should suffer such public humiliation.

The great discovery was not a great discovery after all, and like many promising theories in particle physics at the time, it eventually faded away. But this incident stuck with me. It made me contemplate how dangerous physics might be to one's mental stability.

Developing new theories of physics requires concentrated cerebral focus over long periods of time, and during the course of one's research many failed attempts at solving a problem can lead to frustration. It requires severe discipline to be a creative theoretical physicist. If one is fortunate, one can remain mentally stable through the process. Isaac Newton, however, one of the greatest physicists of all time, was afflicted by periods of madness, and showed anti-social behaviour in his relations with his fellow human beings.

*

The problem of the error in the Einstein-Infeld paper, which had so consumed me at Cambridge during my doctoral work, still nagged at me at RIAS. I wanted to be able to finish that story with a more positive approach to the problem. I sent Banesh Hoffmann a copy of a paper I had recently written, but had not yet sent out for publication, in which I resolved the problem posed by Einstein and Infeld. Hoffmann was a professor at Queen's University in New York, and had been the third author, along with Einstein and Infeld, on the classic paper on the motion of bodies in Einstein's gravity theory, published in the *Annals of Mathematics* in 1938. I wanted to consult with him because this original work that he had been involved with had led to the Einstein-Infeld paper that contained the problem I had found. I received a friendly reply from him, inviting me to visit him in New York. We spent an afternoon at his house discussing the Einstein-Infeld paper. Hoffmann concurred with the conclusions I had reached with Roy Kerr and was enthusiastic about the way I had resolved the problem in the paper I had sent him.

Hoffmann also told me the story behind the original 1938 paper. He had arrived at Princeton as a post-doctoral fellow at the Institute for Advanced Study in the mid-1930s. Einstein had shown him the complicated calculations Infeld had performed on the motion of bodies. Most of the calculations were by Infeld, not by Einstein, who had provided the basic ideas behind the research. Hoffmann discovered errors in the calculations, and Einstein invited him to collaborate on the project. After Infeld and Hoffmann finished the laborious calculations, copies of which are still in safekeeping at the Institute for Advanced Study, they and Einstein submitted the paper to the *Annals of Mathematics* in Princeton. When the galley proofs arrived for final checking, Hoffmann was away, and Einstein never involved himself in tedious work like checking proofs. Unknown to Hoffmann, Infeld sent the corrected galleys back to the journal and changed the order of the authors' names, so that they read "Einstein, Infeld and Hoffmann," rather than the original and more conventional alphabetical order.

Over tea, Hoffmann spent a heated half hour telling me this story; after twenty-two years, he was still angry at Infeld's duplicity. I found Hoffmann to be a wonderful person who displayed much integrity in his scientific research. He was not averse to criticizing Einstein, even though he idolized him. In his later years, Hoffmann wrote a fine biography of Einstein, *Albert Einstein: Creator and Rebel*, with Helen Dukas, who had been Einstein's secretary for almost three decades.

On Hoffmann's suggestion, I submitted my paper to a journal where he was an editor, the *Journal of Rational Mechanics and Analysis* at the University of Indiana, which accepted it for publication. In the paper, now titled "On the Integrability Conditions in the Problem of Motion in General Relativity," I resolved the original problem posed by Einstein and Infeld: the motion of bodies in general relativity. I included a description of the errors in the 1949 paper by Einstein and Infeld, as well as a reference to my earlier, unpublished paper with Roy Kerr.

The senior editor of the journal was a professor in the Physics Department at the University of Indiana, who must have talked about me with his colleagues, for soon I received a letter from Professor Vaclav Hlavaty. A Czechoslovakian immigrant to the United States, Hlavaty had published a book detailing the mathematical intricacies of Einstein's nonsymmetric unified field theory. He invited me to visit Indiana University. Bridget and I went to Bloomington where we met Professor Hlavaty, and I gave a talk on my recent work on the motion of bodies in Einstein's gravitational theory. I also discussed with Hlavaty my correspondence with Einstein and the work that I had done in Copenhagen on Einstein's nonsymmetric unified theory. At the conclusion of our visit, the Physics Department at the university offered me a job as assistant professor.

Upon our return to Baltimore, I evaluated the job offer and what it might mean for my future. I would be back in the academic environment, but I would have to teach courses, which would impede the progress of my research. I finally wrote to Vaclav, thanking him for the offer, but turning it down, explaining that this was only my first year at RIAS, and I hoped to be able to produce more research before taking up an academic position.

*

During my first year in Baltimore, I was invited to attend the first large international relativity meeting, to be held in Chantilly, not far from Paris. This conference was different from the select gathering at the Bern conference; many more physicists were to attend. My immediate association with Chantilly was its famous race course, which was often mentioned in Ernest Hemingway's stories about Paris. Since I had a research contract with the U.S. Air Force Office of Scientific Research, I would be transported to Paris and back, free of charge, on an air force plane.

I stayed overnight at McGuire Air Force Base in New Jersey, joining almost thirty other physicists from all over the United States who also held air force research grants and were being flown to Paris for the conference. At dinner that evening, I met the Turkish physicist Behram Kurşunoğlu for the first time. He had taken a Ph.D. at Fitzwilliam College in Cambridge a year or two before I arrived there, and had achieved some notoriety by having also corresponded with Einstein about his unified field theory. He had published a modified version of Einstein's nonsymmetric unified field theory in the *Physical Review*. But unlike me, Kurşunoğlu had embraced Salam's policy of self-promotion, had spread far and wide the news of his correspondence with Einstein, and also the fact that he had met Einstein at Princeton and discussed his work.

Behram was a flamboyant and assertive personality. At dinner that night, he made the rounds of the tables of physicists, making physics jokes and sly comments about their work. Later in life, Behram put his personality to good use, forming an institute for theoretical physics at the University of Miami and running the annual Coral Gables conferences on particle physics. Making Paul Dirac a member of his institute when he retired to Florida was a significant coup for Behram.

But all of this was in the future. The next day we flew on a military transport prop plane to Paris, stopping off briefly at the Azores to refuel. In the plane, we all sat in seats facing the rear, which was the custom in military planes at the time. During the overnight flight to the Azores, Behram moved from one seat to another, telling amusing stories about famous physicists with great agitation and a subtle edge of contempt.

Another physicist on the flight was Otto Bergman, one of my colleagues at RIAS. After landing at Orly airfield, Otto and I took a taxi to our hotel in downtown Paris. Over dinner that evening at a restaurant in Montparnasse, Otto related how he had been stationed in Paris in the German army. As an Austrian, he had been conscripted in 1943, and had just managed to avoid going to the eastern front. He told me that the German soldiers in Paris had nothing but derision for the French Resistance, and mostly ignored their activities. They considered the Resistance fighters amateurs.

The next day we took a train to Chantilly, and I found myself sharing a hotel room with Wolfgang Kundt, a young, vigorous German relativist, much too young to have had stories to tell of the Nazi occupation of Paris. The meeting was held in a stately château near the town, and there I met André Lichnerowitz, who was the chief organizer of the meeting. He was a professor of theoretical physics and mathematics at the University of Paris, and was one of the best-known relativity physicists in France, having published a book on gravitational theory in French. I was also introduced to several notable relativists, such as Bryce DeWitt, a professor from the University of North Carolina, who was to publish important papers attempting to quantize Einstein's gravity theory.

The ceremonial speech marking the commencement of the meeting took place in one of the large, elegant salons in the château, and afterwards we gathered in a lounge for coffee and snacks. There I met John Wheeler from Princeton, who had kindly directed me to RIAS. We talked about the latest developments in gravitation theory, including the controversial subject of the ultimate fate of the gravitational collapse of stars, with their peculiar singularity properties, which were predicted by Einstein's gravity theory. Several years later, Wheeler would christen them "black holes." An important question then, in 1960, was whether the horizon of the black hole, through which light could not escape, contained a physical singularity or not. In the Schwarzschild solution of Einstein's field equations, a singularity occurs at what is called the Schwarzschild radius at a certain distance from the centre of the black hole. This is a place where Einstein's field equations fail to hold and the laws of physics break down.* I remarked to Wheeler that as of course he knew, Einstein had not been happy with the idea of black holes, and neither had Arthur Eddington at Cambridge. Eddington had disliked the solution so much that he had suggested that there be a law of nature forbidding the existence of black holes.

At this point, a young French waiter floated by, and I asked him for a glass of milk because I was suffering from jet lag and feeling a bit queasy. I drank the milk and continued the discussion with Wheeler. Wheeler had initially been unhappy about black holes, as were most physicists, because of the assumption that the horizon of the black hole was singular. But later, he turned into one of the strongest advocates of black holes. Wheeler explained that neither Einstein nor Eddington had understood the true meaning of

---

*In attacking this problem, Martin Kruskal, a mathematician at Princeton University, and independently the Hungarian mathematical physicist George Szekeres, discovered a mathematical description of what came to be called the black hole solution that proved that the horizon was not a physical singularity. The only true singularity lurked at the centre of the black hole, where the density of matter was infinite.

the event horizon. As proved by Martin Kruskal and George Szekeres, it did not represent a true infinite singularity, but a singularity due to the choice of so-called Schwarzschild coordinates. That is, according to whatever coordinates in spacetime one uses, the result may or may not be a singularity.

All of a sudden, in the middle of this fascinating conversation with Wheeler, I felt faint. I collapsed onto a nearby couch and passed out. I woke up to find myself in a taxi, bouncing along the narrow cobblestone streets of Chantilly. The driver explained that he had been told to take me to my hotel. I managed to pay him and get myself into bed, while feeling that death was surely upon me. I was probably suffering a severe case of salmonella poisoning from having drunk unpasteurized milk, a not-uncommon experience in France in those days.

I lay in bed for two days, feeling almost paralyzed. My roommate, Wolfgang Kundt, came in late at night from the conference, left early in the morning, and was of no help. No doctor came, and no one from the conference checked on me. I finally managed, in a near-delirious state with a fever, to get myself up and out of the hotel to find a pharmacy. The pharmacist gave me a bottle of very large pills, but I misunderstood his instructions in French. Returning to the hotel, I swallowed several of the pills, which unfortunately were actually suppositories. I soon blacked out, and apparently slept for forty-eight hours. I woke up with the most appalling headache I had ever had in my life. I did not manage to attend any of the sessions at the conference. On the last day, I was able to get hold of Otto Bergman, and with my splitting headache we returned to Paris by train. Thus ended my second international physics conference, at which I learned that the hazards of physics are not only mental, emotional and psychological, but can be physical as well!

*

Early on at RIAS, I studied a paper by the German physicist Harry Lehmann at the University of Hamburg. Like many other theorists at the time, he applied quantum field theory to pion-nucleon scattering theory, which predicted what would happen when pi-mesons (or pions, for short) and nucleons collided in high-energy accelerators.*

In his paper Lehmann used a technique in quantum field theory that was called the Källén-Lehmann representation. The problem I had with the paper was that two infinities had to cancel one another at one specific point in spacetime, which seemed to me to produce an inconsistency in quantum field theory. I had written a paper, published in *Nuclear Physics* in 1959, describing this inconsistency. Before submitting the paper for publication, I wrote to Lehmann asking his opinion about this problem, and whether he agreed that there was a serious inconsistency in conventional quantum field theory and in his calculations. Like Infeld two years previously, he never responded to my letter.

My 1959 paper would have repercussions. Once, when I was attending a meeting at Columbia University on particle physics, a tall red-haired young man came up to me after one of the seminars and said, "You're Moffat. I've heard that you've published this paper criticizing Lehmann's calculations in quantum field theory."

"Yes, I have," I answered, sizing up Steven Weinberg, who was a junior member of the Columbia Physics Department.

"You know, I used Lehmann's paper in my lecture notes on quantum field theory that I gave to the graduate students here in the department," he said in an agitated way. "I printed up these

---

*The pi-meson, or pion, in those days before the quark model, was a quantum spin zero particle considered to be an elementary particle carrying the strong nuclear force; it was discovered in 1947 in cosmic rays by Cecil Powell, César Lattes and Giuseppe Occhialini at the University of Bristol in England. "Nucleon" is the generic term for either a proton or a neutron.

lecture notes. I'll give you a copy so you can see for yourself." I saw that he was worried that I was correct, and he had overlooked the importance of this problem.

Word of my paper soon reached Princeton, and back at RIAS I got a phone call from Marvin ("Murph") Goldberger, a professor in the Princeton Physics Department. He also seemed upset about my paper, and explained that he (like Weinberg) had included this work of Lehmann's in his lecture notes for graduate students at Princeton. He invited me to come to the university to discuss the issue. A few days later, I drove up to Princeton and met Goldberger for the first time. We went to lunch at the faculty club, and I found him to be a charming person who treated me with respect and kindness. He suggested that one way to decide the issue raised in my paper would be to meet with Kurt Symanzik, who was currently visiting the Institute for Advanced Study. Would I like to do this?

"Certainly," I said. "It would be interesting to meet Professor Symanzik. The famous paper he wrote on quantum field theory with Lehmann and Zimmermann was impressive." That 1955 paper, "On the Formulation of Quantized Field Theories," had, in fact, contributed to a major revival of physics in Germany, which had suffered severely from the Nazi repression of academia and the war. The paper also represented a significant step in the development of modern quantum field theory.

We walked across the Princeton campus towards the Institute for Advanced Study, down the road that Einstein had taken in his walks between the institute and his house on Mercer Street. At the institute we went into the seminar room where Kurt Symanzik was waiting for us. Goldberger habitually smoked large cigars, and he lit one while we conversed with Symanzik. I found Symanzik's appearance a little overwhelming. He was more than six feet tall and strongly built, with a large, Teutonic head. I knew that he had been one of the last Nazi army officers to be evacuated in the Battle of

Stalingrad before the city fell to the Russians in 1943. In fact, he still looked like the perfect German officer that a Hollywood casting agency would pick for a Second World War film.

Goldberger told us that I should explain to Symanzik the contents of my paper, and we would not be allowed out of the seminar room until one or the other of us was left standing. At this, my stomach began to churn. I looked at the formidable former Wehrmacht officer and realized that any attempt to dominate him was doomed from the beginning.

Goldberger left the room and closed the door. I assumed he was standing outside waiting for the consequences, and wondered why he didn't want to be part of the discussion. Trying not to feel daunted, I went to the blackboard and began explaining my problem with Lehmann's paper.* In particular, he performed a calculation in which infinities occurred at the same point in spacetime, which was physically questionable. I doubted whether this problem could be easily removed, and therefore it seemed to me that this was a potential flaw in quantum field theory.

It wasn't long before Symanzik looked vexed, and we got into a fierce argument about the issue of cancelling infinities. He insisted on his point of view, claiming that quantum field theory was quite self-consistent—he didn't even consider the infinities to be a problem—while I adopted the opposite point of view, claiming that there was a deep underlying problem with the whole renormalization scheme, and that hopefully someday quantum field theory would be reformulated in a more consistent and rigorous fashion. I pointed out that one of the inventors of quantum field theory, Paul Dirac, had never believed that the renormalization scheme

---

*The problem had to do with the technique of renormalization in quantum field theory. Renormalization is a mathematical way of absorbing certain ugly infinities in the mass and charge of a particle in such a way that the final result only depends on the measured mass and charge.

in quantum field theory was acceptable as a physical theory. However, it had become the standard way of solving quantum electrodynamics, the theory that quantizes Maxwell's electromagnetic field equations in a way that is consistent with relativity theory. Although quantum electrodynamics was in startling agreement with all the observations, it still, in my opinion, lacked a fundamental, rigorous formulation.

Thanks to my experience with Wolfgang Pauli as a student at Cambridge, I had learned to withstand the severe attacks of older, renowned physicists on my research, and I persevered with Symanzik. Finally, I could feel that he was losing interest in the discussion; he declared impatiently that the session was over. It seemed to me that the whole issue of the inconsistency of renormalization theory remained unresolved, and our debate was a draw. Both of us were left standing. We left the seminar room together and found Goldberger pacing outside, puffing on his cigar.

The issue as to whether quantum field theory is a physically consistent theory is still controversial today, fifty years later. I myself am still trying to develop a more satisfactory quantum field theory. Many physicists are searching for a physically consistent quantum gravity today. Quantum gravity is an attempt to quantize Einstein's gravitational theory, which can take the form of quantizing the geometry of spacetime. As with Maxwell's classical theory of electromagnetism, Einstein's classical gravity theory should allow a quantum description. Much of the lack of success of current quantum gravity research is due to problems with quantum field theory. The advent of string theory was thought to make quantum field theory, as it was originally formulated, less relevant. But with the lack of predictive tests of string theory, many in the theoretical physics community are returning to attempts to make quantum field theory more rigorous.

In 1960, in Baltimore, with all the physics activity going on in my life, our first child, Sandra, was born in January. Since she was born in Baltimore, she automatically became a U.S. citizen. At the same time, my increasing interest in particle physics led me to request a leave of absence of several months from RIAS so that I could go overseas and work at the large accelerator facility at CERN, in Geneva, Switzerland.

13

CERN AND

PARTICLE PHYSICS

*I*N THE SPRING of 1960, I become a visiting fellow in the Theory Division at CERN. CERN—Conseil Européen pour la Recherche Nucléaire—situated just west of Geneva, and overlooked by the dramatic Jura Mountains, was the centre for the world's largest particle accelerator, built in 1957. A theoretical physics group had been established in rather ill-constructed buildings adjacent to the laboratory, where the larger group of experimental physicists worked. There was a huge effort at CERN in those days to study the properties of the new particles that were discovered almost every month in the proton-proton collider, and there was a great deal of excitement about all the anticipated discoveries that would shed new light on the subatomic structure of matter. Particle physics thrives on experimental data and, when the machine started up, it was a big boost for particle physics theorists to promote new ideas about what the accelerator was finding.

The social centre at CERN was, and still is, a large cafeteria with an outdoor terrace where physicists gathered for coffee and lunch and discussed the latest developments in particle physics. It was a cosmopolitan group, with physicists from all over the world, including, at the height of the cold war, some visiting theorists from the Soviet Union, which had built a competing accelerator laboratory at Dubna, close to the Volga River. A pool of secretaries was attached to the theory division, and CERN had its own bank in the main foyer, and a good physics library. I found it an ideal environment in which to work.

While at CERN, I managed a side visit to the Niels Bohr Institute in Copenhagen, the first of many such visits to the institute over the years. I was disappointed not to see Niels Bohr himself, for he had retired. On that first visit, I met Gunnar Källén in person, the other member of the Källén-Lehmann representation, which had played such an important part in my 1959 particle physics paper. Källén was a professor at the University of Lund in Sweden and was also visiting Copenhagen to participate in seminars and discussions. He invited me to give a talk at Lund about my paper on quantum field theory. Lund was an easy trip from Copenhagen, and I accepted the invitation. I had heard the gossip among physicists that Källén and Lehmann were embroiled in a dispute over who should get credit for the Källén-Lehmann representation. Consequently, I surmised that Källén was not unhappy about my finding a problem in one of Lehmann's papers. I stayed in a modest hotel in Lund near the campus. Walking around the town, I found that its small-town atmosphere mingled charmingly with the atmosphere at Lund University, which was one of the oldest universities in Europe.

Källén had been a student of Wolfgang Pauli in Zurich, and had published papers that contributed to the development of quantum electrodynamics. At the time I was in Lund, Källén was still a young man, only about thirty-six years old, and he had already published an important book on particle physics. With his light blond hair and striking blue eyes, he looked quintessentially Swedish.

I gave my talk to the assembled group of professors and students, and only a few minutes into my scratching on the blackboard, Källén interrupted. I began to feel the shadow of Pauli falling over the whole proceeding. As with Pauli, Källén continuously interrupted me with aggressive questions, even though he was not disagreeing with my technical criticisms of Lehmann's paper. Rather, his interruptions seemed simply to be his style of doing physics. After half an hour of this, I felt like striding over to him, handing him the chalk and saying, "Professor, why don't you give the talk?" However, I managed to control myself and finished without getting too riled up. At that time it was the tradition in European academic settings to interrupt the speaker in an aggressive way. This sometimes made it difficult for the speaker to ever get his point across. During the seminar, none of the students said a word, nor did the other professors in the audience. Källén dominated the whole session.

Källén had invited me to his home for dinner on the evening after my talk, and it was with some trepidation, after the grilling I'd suffered at the seminar, that I made my way to his house not far from the university. Yet the evening could not have been more pleasant, nor the professor more charming. It was as if he had one persona for the seminar room and another for his role as host. The Källéns' magnificent Swedish-style house was spacious and orderly. Källén's wife was a lovely blond woman, younger than he. The dinner proceeded in a formal way. At the head of the table, Källén had a small contraption next to his plate with a button; occasionally he would press it and a maid would appear in a black uniform and white lace hat to serve the courses and pour the wine. Several years after this dinner, in 1968, I was shocked to hear that both Källén and his wife had been killed when their small private airplane crashed in Denmark.

*

Particle accelerators come in two types, either circular beams of particles such as protons and anti-protons hitting one another or hitting a fixed nuclear target; or linear accelerators, which use beams of electrons and positrons. Strong electric fields accelerate the particles in the beam to relativistic speeds close to the speed of light, while magnetic fields focus them. The first CERN accelerator, built in 1957, was the circular type, called a cyclotron, which reached an energy of 600 million electron volts.* In 1959, CERN built the proton synchrotron, which for a brief time was the highest-energy accelerator in the world, accelerating protons to an energy of 28 billion electron volts. Physicists study the debris from the high-energy collisions to investigate the inner workings of the atomic nucleus.

During my time at CERN, I worked exclusively on particle physics. While at RIAS, I had published a paper in the *Physical Review* developing what I called the inverse scattering amplitude method. The paper described a novel way of predicting what the cross-sections—the areas that particles scatter through when they collide in an accelerator—would look like. When experimentalists performed a collision experiment, their results came in the form of numbers of particle collisions as a function of energy. They needed theorists to interpret the numbers so that they could understand the details of the particle collisions.

My method was a new way of predicting what particles do when they collide. In particular, I was concentrating on the scattering of two pi-mesons, which are massive, unstable, elementary particles with a quantum spin zero, that are produced in the accelerator through particle collisions. The problem was to explain theoretically what was happening when two pi-mesons collided. At the time

---

*An electron volt (eV) is the kinetic energy that an electron gains as it is accelerated between two conductors with a one-volt difference in electric potential. According to Einstein's famous equation that relates mass to energy, the mass of a particle can be expressed in electron volts.

physicists believed that these pions carried part of the strong nuclear force that held protons and neutrons together in the atomic nucleus.

My research fitted in with the worldwide project of the S-matrix, invented by Werner Heisenberg in the 1940s as a way to describe the scattering of particles in the accelerators at the time, which were much smaller. Quantum field theory had taken on a far less significant role in particle physics, and the subject of dispersion relations and S-matrix theory dominated the field in the early 1960s. Two influential figures in this subject were Geoffrey Chew and Stanley Mandelstam at Berkeley, California. At CERN, there was a competing group led by Sergio Fubini and Daniele Amati, who were also attempting to describe pi-meson (pion) scattering.

Brian Bransden, another research fellow from England who was at CERN for one year, had an office near mine in the Theory Division. Brian, though only a few years older than I, was a world authority on atomic physics and had published several important papers on this subject. He had come to CERN to focus on particle physics and in particular to better interpret the scattering of particles in colliders.

Over coffee one morning we discussed my method of interpreting the collisions of particles and applying it to pion-pion scattering. Both my method and the well-known method of the Berkeley group* had the important property of guaranteeing what was called the "unitarity of the S-matrix," which means that it ensured that the sum of the probabilities of the various outcomes of the scatterings always added up to exactly 100 per cent. I wanted to solve the problem using my own published techniques, and persuaded Brian to

---

*The method used to solve the problem of pion-pion scattering by the Berkeley group was based on what was called the Mandelstam representation, and what was figuratively called the "N over D" method (meaning, literally, the numerator divided by the denominator method).

collaborate with me. He was highly proficient at solving problems in atomic physics with computers, and we could use his techniques to apply my methods to the scattering of pions.

The main computer at CERN for theorists in the early 1960s was a Ferranti Mercury, which filled up a room the size of a small ballroom. It consisted of a vast array of vacuum tubes and spools of magnetic tapes. The data to be computed was transferred from punch cards read by a special card machine. The computer performed the computations overnight, and an assistant handed out the results, printed on large stacks of paper, to the physicists the next morning.

This was the very beginning of the computer era in science. At the time, we felt a sense of new power, being able to harness and use the incredible computational capacity of this huge technological beast. Today it is humbling to realize that the power of the 1960 CERN computer was roughly *one millionth* the power of an ordinary laptop available today. Sometimes we ran computations on the enormous Ferranti Mercury for days, while today those same operations would take seconds on a laptop.

Brian Bransden and I met every morning in the computer room where the overnight runs were collected, and pored over the results. After a few months of this arduous effort, we discovered a very exciting result. In our calculations, we discovered what was called a resonance in the P-wave scattering channel for pion-pion scattering.* In other words, we were predicting the existence of a new particle, which had not yet been experimentally observed! This new particle, which we theorized would leave its unique signature in the debris after a collision, had a quantum spin of one.

---

*The scattering amplitudes for particles were projected into angular momentum states named after the spectroscopic states in atomic physics, namely, S, P, D . . . waves.

Particles discovered in the accelerators were unstable, with a lifetime of fractions of seconds, after which they decayed into other particles with smaller masses. These highly unstable particles were detected as resonances in collider experiments; that is, excited states when two particles would reinforce one another at certain energies. These resonances revealed themselves in scattering cross-sections as "bumps," and the width of the bump indicated the lifetime of the particle. Our computer calculations predicted the existence of such a bump at around 700 MeV (million electron volts), with a very short lifetime. We named our new particle the P-wave resonance.

Brian and I were excited about this result: from our theoretical calculations, we had predicted the existence of a new particle, which the experimentalists could now set about finding in the accelerator collision data. But we were cautious about our prediction too, and we began considering the best way to check our calculations. As Brian was an old hand at applying computers to physics, he announced one morning that we would have to check every single line of the computer program, to make sure there were no mistakes in our calculations. This meant checking hundreds of computer code lines by hand. We each got hold of a hand calculator and, taking up positions at opposite ends of my office, we spent days going through the huge computer program, each of us independently checking it line by line, with thick stacks of computer output next to our calculators. Finally, after weeks of this painstaking work, we concluded that the predicted resonance bump was real.

I gave a seminar in the Theory Division about our results, which were received with much interest by the theorists and experimentalists. The prediction of a new particle was a significant event. As was usually the case at these theory seminars, Claude Lovelace, a visiting fellow from England, sat in the front row with a huge pile of computer printout paper on the table in front of him, the results of his computations of the night before. While I spoke about our exciting results, he leafed through the printouts, making continuous, loud rustling noises, and paying no attention to what I was saying. Lovelace did this habitually, no matter who was speaking. Regardless of the weather, Claude was always attired in shorts and leather sandals, most often without socks. His long greying beard and massive thatch of hair gave the impression of a miniature Sasquatch man.

Later in the year, Lovelace was scheduled to give a seminar to the theory group. The day before the seminar, a physicist prone to pranks and hoaxes proposed that every member of the audience at Lovelace's seminar should bring along a large stack of printout paper, and at a given signal, we should all start leafing through them and pay no attention to Lovelace. And so, the next day we all brought along our stacks of computer printouts and waited for the signal from the prankster. Lovelace was busy speaking and writing on the blackboard with his back turned to the audience. When the signal was given, we all began noisily leafing through our printouts and carefully studying the computer calculations, ignoring Lovelace's sonorous voice speaking about his results on scattering theory. After a while I looked up and saw that Lovelace had turned around, and was staring mutely at the audience with his mouth open. He was not amused, but the rest of us suddenly erupted into laughter. The lesson, however, was lost on Lovelace, for at the next theory seminar he turned up again with his standard computer printouts and went through his usual ritual of rustling through the stack, paying no attention to the speaker.

Sometime after my talk to the group, Brian and I decided to write up our results on the P-wave pion resonance as a preprint to send out from CERN. A secretary in the Theory Department's pool typed the paper and duplicated it on a Gestetner machine. This was a laborious task before the invention of word processing, involving endless revisions, necessitating new Gestetner forms and corrections to the old ones. We had to face the tight-lipped secretary whenever we presented a new revision to her. Brian and I also had to produce a figure showing the predicted bump in the cross-section for pion-pion scattering, and this had to be drawn with a metal pen on the plastic sheet used to print the pages on the Gestetner machine. I was the one delegated to draw the bump. In spite of my background as an artist, it turned out that I was not very good at this. My metal pen slipped and made a small kink in the bump, which would have serious future consequences.

All papers submitted for publication had to be approved by two or three members of the Theory Division, including the director, who at that time was Leon van Hove. Van Hove was a humourless Dutchman who had established his reputation with a well-known paper on quantum field theory. One of the visiting members of the theory group at the time was Gordon Feldman, my friend from Johns Hopkins in Baltimore. When I showed Gordon our result, he laughed and said, "Why are you publishing this paper claiming that this particle exists, when no one has ever seen such a particle resonance in pion-pion scattering?"

"I think this is important because it constitutes a theoretical prediction for this particle," I replied. "Hopefully, someone will see it soon in an accelerator."

Feldman looked at me doubtfully and said, "Well, I guess it's not going to do too much harm to put out this paper."

The competing Chew-Mandelstam group in Berkeley had put out a paper recently showing results of pion-pion scattering using Mandelstam's technique to solve the S-matrix problem, but they had concentrated on the S-wave scattering, and did not demonstrate explicitly the existence of a P-wave resonance, as we had done. We submitted our paper to *Physical Review Letters* and it was published on June 15, 1961. The paper had previously been issued as a preprint from CERN in December, 1960. A longer version was published later in 1961 in the European physics journal *Il Nuovo Cimento*, and in 1962, we published a follow-up letter in *Physical Review Letters*, describing the final solution for the P-wave resonance prediction.

During this time, when Brian and I were busy with our computations, Bridget and one-year-old Sandra were left behind at a small chalet-type house just above the highway skirting Lake Geneva. I would often stay over at CERN, because we would work far into the night. Nominally, the CERN computing division was closed at night while the machines hummed away at their calculations. One night I was desperate to find out the result of a particular calculation, so I went into the grounds of CERN, pried open a window into the computing room, and slid through and dropped down to the floor uninjured. I retrieved a magnetic tape and managed to copy the results of our calculation onto it and printed the result myself. I was quite proud of this achievement, and when I told Brian about it the next day, he laughed.

However, there were weekends when I took my family into the mountains in our small Volkswagen Beetle. I took my skis along, and while Bridget and Sandra amused themselves in a restaurant, I would repeatedly go up the mountain in a chairlift to ski back down. One time we visited the town of Chamonix below Mont Blanc, which rose up as one of the tallest mountains in Europe above us, casting a shadow on the village below. Foolishly, I stayed up on Mont Blanc by myself until late in the afternoon, and discovered that as the sun sank, the temperature dropped rapidly, forming ice on the ski run and making it dangerous to descend. I was one of the last skiers stuck up there on that late afternoon. Fortunately, three French alpine ski patrolmen arrived and helped me gingerly find my way down the mountain. Bridget displayed laudable patience during this humiliating display of macho activity.

As it turned out, the existence of a pion-pion resonance in the P-wave, or a short-lived particle, was confirmed in 1961 by an experimental group led by W.D. Walker at the University of Wisconsin. They named it the "rho-resonance." Its mass was 760 MeV, close to the mass of 700 MeV that Brian and I had predicted. The resonance was unstable and decayed into two pi-mesons, just as we had predicted. The paper was published in *Physical Review Letters* on June 1, 1961, just two weeks before our paper came out. The prediction of the rho-meson, or our P-wave resonance, had been a highly competitive enterprise. I would learn later, upon my return to the States, that one of the leading figures in the field had been unhappy about our publishing this work, and thereby scooping him.

I was dismayed that Brian Bransden and I did not get more credit for predicting the existence of this particle, which played an important role in the future development of particle physics.

Yet, perhaps on the strength of my work on pion-pion scattering, I was offered a position at CERN. This was a lucrative position, for foreign national permanent members at CERN did not pay Swiss income tax. But I felt an obligation to return to RIAS—and Bridget also found living in a non-English-speaking country difficult with a child—so I turned down the offer. I consider this one of the worst career mistakes in my life. A position at CERN would have given me the opportunity to concentrate on my research unencumbered by the heavy teaching duties in academia. Moreover, the CERN Theory Division was a focal point for physics research in the world in the 1960s and is perhaps even more so today, as experiments at the Large Hadron Collider (LHC) begin.

Only a couple of weeks after our return to Baltimore, in January 1961, while I was attending an annual meeting of the American Physical Society in New York, I received a phone call from Bridget informing me that my father had had a heart attack and died in Cambridge. He was only fifty-three, so this was a terrible shock to me. I returned to Baltimore immediately, and then went back to

New York and took an overnight flight to England. My mother was devastated, and after my father's funeral, she agreed to return with me to Baltimore, where she lived with us for six months before moving back to Denmark.

14

PRINCETON AND

OPPENHEIMER

W HEN I WAS SETTLED at RIAS again after my time at CERN and my father's funeral, Robert Oppenheimer invited me to give a talk at the Palmer Laboratory in Princeton. Oppenheimer, whom I had met previously at the Einstein Fest in Bern, was one of the gods of U.S. physics in the midtwentieth century. He had been the director of the Manhattan Project that produced the first atomic bomb. He was interested in the preprint that Brian Bransden and I had sent out from CERN on our prediction of a resonance in the P-wave pion-pion scattering amplitude, later known as the rho-meson. This was still months before the actual publication of the paper in *Physical Review Letters*.

At this time, I was beginning to feel apprehensive about the future of RIAS, and I thought it was time to put out feelers for an academic position. Welcome Bender, our director, who was not a scientist, officially worked at the Martin aircraft company even though he was director of our institute, where only fundamental physics and mathematics research was being carried out. During this time of the cold war, the army, navy and air force, which supported our fundamental research, always believed that we would produce the science that would lead to a useful new lethal weapon to threaten the Soviet Union. Perhaps my research on pion-pion scattering suggested to the military that I might be able to produce a "pion bomb." On the other hand, I had heard rumours that the aircraft company had lost enthusiasm for supporting a fundamental-research institute, and it was possible that they might close it down in the near future, which made me feel even worse about having turned down the job offer from CERN.

Thus, the invitation from Oppenheimer to speak at Princeton sounded like a good first step towards applying for a position at Princeton or elsewhere, and perhaps getting a letter of recommendation from Oppenheimer.

My talk at Princeton was actually organized by the celebrated quantum field theorist Arthur Wightman, who was a professor at the Palmer Laboratory, which was the Physics Department at Princeton University. He invited me to stay at his house in Princeton, and we had a delightful evening together, as he and his wife were charming hosts. Wightman was a tall, handsome man in his thirties, and his wife was a tall, striking woman who also had a faculty appointment at Princeton. During dinner, the conversation flew along on erudite topics, such as the fine points of Greek mythology and many arcane details about the tsetse fly, which causes so many deaths in Africa through sleeping sickness.

I gave my lecture on a Thursday morning before lunch. The title of the talk was "New Results on Pion-Pion Scattering and the Inverse Scattering Amplitude Method." Already a little nervous before the lecture, when I walked into the seminar room, I was taken aback at the number of famous physicists who sat staring at me as I approached the podium. Oppenheimer, of course, was there, as well as Murph Goldberger, Samuel Treiman and Eugene Wigner.

Wightman introduced me as a young physicist who appeared to have solved some fundamental problems in pion-pion scattering. After a nervous beginning, I got warmed up, and gave a detailed presentation on the discovery of the P-wave resonance that Brian and I had made during the previous year at CERN. However, soon, members of the audience, particularly Sam Treiman, a well-known professor in particle physics, started bombarding me with intense questions. I tried to keep calm and answer his questions as well as possible. When I finished my talk, the formal question period began, and I was assailed by yet more questions. It was all reminiscent of my experience with Pauli, and then his student Källén, and I was glad I had learned to remain relatively unrattled when under attack by older physicists.

Wightman came up to me at the end of the talk to tell me that we were going to lunch prior to a faculty meeting in the Physics Department. The lunch was held in a faculty room, and I sat opposite Eugene Wigner, with Arthur Wightman on one side of me and Valentine Bargmann, the renowned Princeton mathematician, on the other. Murph Goldberger presided at the head of the table, and presumably was going to chair the faculty meeting after lunch too. Eugene Wigner, Paul Dirac's famous brother-in-law, was a Nobel Prize winner for the important work he had done in nuclear physics. He was well known for his polite, even obsequious demeanour and for asking simple questions that could turn out to be devastating. He leaned across the table and asked innocently, "Moffat, do you mind if I ask you a simple question? Can you explain to me what crossing symmetry is?"

Wightman chuckled, and Goldberger also seemed amused. I explained as well as I could what I understood about crossing symmetry, a fundamental mathematical property of scattering theory. Wigner peered at me intently for a few moments, said nothing further and continued eating his lunch. I sat and wondered whether my future would depend upon how I had answered Wigner's question.

*

The next day, before returning to Baltimore, I planned to visit the Institute for Advanced Study, which was some distance away from the Princeton campus. I wanted to let Professor Oppenheimer know that I was interested in an academic position, to sound out whether anything might be available at Princeton or the institute, and to ask whether he would be willing to write a letter of recommendation for me. I walked through the hallowed halls of the institute, found a door bearing a plaque that said "J.R. Oppenheimer," knocked and walked in. A stern-looking, middle-aged lady sat at a desk. Her grey-ing hair was combed back tightly into a bun and she wore thick-framed glasses. "Can I help you?" she asked.

"My name is John Moffat. I'm from a research institute in Balti-more. Professor Oppenheimer invited me to Princeton to give a talk, which I gave yesterday."

She said curtly, "Yes, I heard about this."

"Is it possible for me to see Professor Oppenheimer before I return home?" She rose from her desk, walked over to another door and knocked on it. There was a loud "Come in!" from inside.

Several minutes passed while I waited for the secretary to return from Oppenheimer's inner sanctum. Finally, she came out, closed the door behind her and said brusquely, "I'm sorry. Professor Oppenheimer cannot see you." I got up, feeling baffled, and didn't know what to say. I thanked her lamely and left.

As I wandered around the institute grounds, I pondered why Oppenheimer wouldn't see me. Was my talk that bad yesterday? I didn't think so. Did I fail some secret test at lunch, which he had heard about? What had I done wrong? So much for asking him for a letter of recommendation in my applications for university positions. Such was academia. You could never be quite sure about your position. The behaviour of academicians, and physicists in particular, tended to encourage a certain degree of paranoia in one's personality.

Some months later, I ran into a colleague at a physics conference who had heard about my talk at Princeton, and who was doing research at the time in Berkeley. He told me that he had heard via the grapevine in the Physics Department there that one of the prominent professors in the S-matrix group had heard about Oppenheimer inviting me to give a talk at Princeton, and, being upset with Bransden and me for beating him to the prediction of the P-wave resonance, or rho-meson, had sent a telegram—the e-mail of the 1960s—to Oppenheimer. In it, he said that he had read the preprint sent out from CERN by Bransden and myself, and noticed right away that the figure describing the resonance bump in the P-wave scattering amplitude had a kink in it. This kink could only be caused by a serious discontinuity in the calculations we had performed, the telegram continued. Consequently, the whole calculation must be incorrect, and Oppenheimer should disregard everything I said.

I was astounded by this report, suddenly recalling my difficulties with the Gestetner instrument, and demanded to know from my colleague whether he really believed this was true. He said yes, he was pretty certain it was correct. I already knew that physicists could be brutal in their treatment of competitors, but this was truly beyond belief. Of course, it explained why Oppenheimer had refused to see me the day after my talk, and it certainly killed any hopes I had had for obtaining a position at Princeton.

*

Despite this dispiriting experience, I began travelling to Princeton every other week to attend seminars at the Institute for Advanced Study. It was a wonderful opportunity to hear well-known and talented physicists such as Abraham Pais, Julian Schwinger, Murray Gell-Mann and many prominent visitors talk about particle physics and quantum field theory. As director of the institute, Oppenheimer presided over the lecture, and discussions continued over afternoon tea.

Oppenheimer was in the habit of dominating the seminars. I learned that he would spend an hour or two preparing for a guest speaker's seminar, and then would usually sit in the front row, close to the speaker in the institute's moderately sized seminar room. There were usually no more than fifteen to twenty physicists in attendance. Not long after the speaker had commenced, Oppenheimer would almost invariably interrupt, and explain to us what the speaker was really trying to say.

John Geoffrey (J.G.) Taylor, who had been a student at Cambridge during my years there, was visiting the institute as a fellow, working on the dispersion relation methods in particle physics. He told me that on arriving at the institute, he and his wife, Pat, were invited to the Oppenheimers' impressive house on the outskirts of Princeton. The evening had turned out to be embarrassing. At the outset, Oppenheimer's wife had offered them drinks, and unfortunately, Pat dropped her glass of red wine on the floor on an expensive rug. Oppenheimer rushed out of the room, returning with a pail of water, which he threw dramatically on the rug in front of them. Later, in a conversation after dinner, Pat Taylor intimated that she had learned to play the piano as a child. The Oppenheimers insisted that she play their Steinway grand piano, and a nervous Pat Taylor was forced to play Chopin. Not having practised recently, she did not perform very well, much to the Taylors' embarrassment. Oppenheimer, in his usual critical way, made a testy comment about her playing.

John Taylor and I were in a prankster mood one day, and agreed it was time for "Oppie" to learn a lesson. Besides, I had never gotten over Oppenheimer's treatment of me after my talk at Princeton. At Oppenheimer's seminar that day, an invited speaker was talking about dispersion relations, which both Taylor and I were working on. Sure enough, after no doubt preparing himself in his office to do battle at the lecture, Oppenheimer sat close to the speaker and rudely interrupted him. Taylor and I were stationed at opposite sides of the room, watching each other. On a signal from me, Taylor interrupted Oppenheimer while he was in the process of haranguing the speaker, and asked him a technical question that was not easy to answer. Oppie looked momentarily fazed, but then ignored Taylor and returned to dressing down the speaker. I then interrupted and asked Oppie another highly technical question, following up on Taylor's question. Oppie looked at me quizzically, with an irritated frown, and tried to answer my question. "Well, Professor Oppenheimer, this is not the correct answer," I said matter-of-factly. From the other side of the room, Taylor butted in and agreed that what Oppie was saying was absolutely wrong.

At this point, Oppenheimer appeared to realize that we were ganging up on him. Suddenly he stood up, white in the face, strode across the room, opened the seminar door and slammed it shut as he exited.

Taylor and I suffered no repercussions following this seminar, but it was obvious that our little lesson had been completely lost on Oppie. At the next seminar, given by his former student Robert Serber, then a professor at New York University, Oppenheimer behaved in his usual egregious fashion, interrupting Serber relentlessly. At one point, he stopped Serber in midsentence, turned to the audience and said, "This is what Robert is trying to say," and gave a five-minute summary of his talk, while Serber looked on helplessly.

Despite Oppenheimer's critical and rude behaviour, the Princeton seminars were delightful. The talks were informative and I learned a great deal about theoretical particle physics.

Later, when I recalled the incident with Taylor and myself at the seminar, I felt some regret about our behaviour towards Oppenheimer, who, despite his autocratic personality, was highly respected in the United States, as well as internationally. One understood from accounts by people on his Manhattan Project team that he had been a very successful administrator; indeed, his leadership was in large part responsible for the success of the project. When Oppenheimer later described the first test of the atomic bomb in the New Mexico desert, he made his famous statement about the grave responsibility of having produced this devastating weapon, quoting Hindu scripture:

A few people laughed, a few people cried, most people were silent. There floated through my mind a line from the Bhagavad-Gita in which Krishna is trying to persuade the Prince that he should do his duty: "I am become death: the destroyer of worlds."

In later years, Oppenheimer opposed Edward Teller's program to produce a hydrogen fusion bomb. During the McCarthy era, he was investigated for his involvement with the Communist Party while a young physicist at U.C. Berkeley. I think that the subsequent withdrawal of his security clearance by the FBI had embittered him, and might partially explain his behaviour at the institute seminars. It was clear that Oppenheimer was a complex person. I learned later that while he was a student at Cambridge, England, doing laboratory research, he had become insecure about his work, and had actually attempted to poison his supervisor, Patrick Blackett. Remarkably, Cambridge University authorities treated him with leniency and allowed him to finish his research under close supervision.

Oppenheimer's bizarre behaviour to me, personally, reasserted itself later when I encountered him at a physics conference. As I approached him and put out my hand, he said, smiling mischievously, "Ah, Bransden!" I said hesitantly, "No, Professor Oppenheimer, I'm Moffat." This exact exchange happened again not long afterwards, at another physics meeting. And again, Oppie smiled and seemed amused by his own trickery.

*

Our years in Baltimore coincided with the civil rights movement in the United States and with the cold war. Once, in October 1962, when I was driving from Baltimore to Princeton to attend a seminar and stay overnight, I caught the news on the car radio of the imminent launching of a nuclear warhead from Cuba. This was the climactic day of the Cuban missile crisis, and John F. Kennedy and Nikita Khrushchev were butting heads. The nuclear warheads were predicted to land on the east coast of America, pretty much where I was driving at that moment. I thought about Bridget and two-year-old Sandra back in Baltimore, and for several minutes debated with myself about turning around and going home. However, I continued to Princeton, and before my drive ended, President Kennedy had managed to prevent the crisis from escalating into war.

Just over a year later, I attended a conference in Dallas on new developments in astrophysics. It was only two days after the assassination of Kennedy. At the meeting's opening ceremony, the organizers asked the audience for a three-minute silence in memory of President Kennedy. One day I went to visit Deal Plaza, and stood beneath the window of the book repository building from which Lee Harvey Oswald had shot Kennedy. The whole plaza and even the boulevard leading into it were carpeted with bouquets of flowers. On an American Airlines flight returning to Baltimore, I sat next to a large, stocky Texan in cowboy hat and boots. During a brief discussion with him, he told me that he was happy that someone had finally killed that "son of a bitch," Kennedy.

The Dallas meeting had been an important one for the field of astrophysics. New observational results on quasi-stellar sources, later called "quasars," were announced. This was a subject of much investigation at the time: How could such relatively small objects produce such large fluxes of energy? And how far away were they from us? I was fascinated by the new-found quasars, and soon made them a subject of investigation at RIAS.

But it was the aftermath of the Kennedy assassination, and the mixed feelings of grief and perverse satisfaction among the Texans I met, that remained with me after the Dallas conference. As a transplanted British citizen, I sometimes felt out of my depth in this often-violent country that also nurtured many of the most talented leading physicists in the world.

15

TORONTO

I N THE SUMMER of 1962, I was invited to Hamilton, Ontario, to give a talk on my current work in particle physics at the annual meeting of the Canadian Association of Physics. I took a train from Baltimore to Hamilton, and spent a week there at the conference. Robert Marshak, who was then chair of the Physics Department at Rochester University, arrived at the meeting an hour before my talk to chair the session, and I was impressed that soon after the session, he took off in a taxi to return to Rochester. This contrasted with my week-long stay, which surely made it appear as if I had nothing more important to do with my time. I hadn't learned how to create the image of being an important and busy physicist. I regret to say that after fifty years as a physicist and professor, I still haven't learned how to play this game.

The next year, in 1963, a letter arrived from Harry Welsh, chairman of the Physics Department at the University of Toronto (U of T), inviting me to apply for a position there, and to visit the university and give a lecture. He had heard from U of T physicists who had attended the conference in Hamilton that I might be a good addition to the Physics Department. This time I flew from Baltimore to Toronto via New York. There was no particle physics group in the University of Toronto Physics Department at that time, and Professor Dick Steenberg, a nuclear theorist, wanted to establish one. In fact, he and Harry Welsh had in mind that I could start a particle physics group.

Soon after my return to Baltimore, I received a second letter from Harry Welsh, offering me a position as associate professor with tenure. This was quite unusual, for the usual academic route begins with the position of assistant professor without tenure, and then typically, after five years, one would be considered for promotion to associate professor with tenure, and then after several more years, one could entertain the possibility of being promoted to full professor. So I was able to short-circuit five years of waiting for tenure. Probably one of the reasons for this generous offer was that I was going through a very creative period in my physics, and had published more than a dozen particle physics papers in the *Physical Review* during 1961–63 on what was then cutting-edge research in particle scattering theory, S-matrix theory and Regge poles. After some serious deliberation and discussions with Bridget, and eager to have a more secure position to support my family, I accepted the Toronto offer.

In July 1964, we drove from Baltimore to Toronto in my old Jaguar. Bridget and I rented an apartment in a rather uninspiring suburb on the west side of the city, near the airport. I was faced with the prospect of having to give lectures starting in September. I had never taught a course to students or had any teaching duties until then. In fact, I had never even officially taken a course at university! At Cambridge, I had sat in on two courses, but never turned in assignments or took an exam. I was apprehensive about my new role as a teaching professor. That first year I had to teach a quantum mechanics course to engineers and a graduate course on particle physics and quantum field theory. In addition, I had to give two-hour tutorials on first-year physics in the basement of the old mining building attached to the Physics and Engineering departments starting at eight in the morning. Never at my best early in the day, this tutorial was what I dreaded the most about my new job. In addition to teaching, I needed to keep up my full-time research; I intended to continue producing a large crop of papers. My first year in Toronto was a stressful one, but I managed somehow to give a reasonable account of myself as a teacher and to publish several more particle physics papers.

*

Over the next few years, I built up a particle physics group at U of T, was promoted to full professor in 1967—the same year that our second child, Christina, was born—acquired talented graduate students and survived the turbulent 1960s on a large North American campus. In 1968, when I was thirty-six, Harry Welsh, still chair of the department, asked me to join a three-person committee to create a new physics curriculum. This turned out to be an arduous task, and put extra strain on me as I continued teaching my courses, doing my research and supervising graduate students. But the new curriculum was a success, and remained the model for the core undergraduate teaching in the Physics Department for decades. I considered it ironic that, despite my never having taken a physics course for credit in my life, I was able to help formulate this successful undergraduate teaching program at what was now one of the major physics departments in Canada.

In my research during those early years in Toronto, I continued working on S-matrix theory and other aspects of particle physics such as the then-popular Regge poles, alone and with my graduate students.

*

In 1972, I was eligible to take a sabbatical leave, and I chose to go back to Cambridge. I had been away from Cambridge for fourteen years, and having now achieved a respectable position in academia and some renown as a physicist, perhaps subconsciously I desired to relive my student days. The leave was to last a year. We lived in a rented flat, and our older daughter, Sandra, attended school in Cambridge.

During that sabbatical year, I was invited to a winter school on particle physics held at a ski resort in Schladming, Austria. I looked forward to this eagerly, as a chance to combine physics with my favourite sport. The physics part of the week consisted of a course of lectures on S-matrix, Regge poles and scattering theory. Since I was considered an expert on these topics, the organizers of the school had invited me to give talks about them.

At the first morning of my lectures, which began at one of my least favourite times of the day, eight-thirty in the morning, a stocky young German student, Bruno Renner, approached me as I was standing at the podium ready to begin. I had eaten some Wiener Schnitzel the evening before that had upset my stomach, and I was feeling a little off colour. Renner looked up at me and said in a German accent, "Professor Moffat, why are you giving these lectures on S-matrix and Regge poles? We students are no longer interested in this subject. We're now doing quantum field theory and current algebra." The latter subject was closely related to quantum field theory and group theory, and was being promoted by Murray Gell-Mann.

I looked down at Renner morosely and thought, "This is a great beginning to a week's lectures." As it turned out, the sentiments of the students at the physics-ski school correctly predicted the course of particle physics during the 1970s, when quantum field theory took over, and then the next dominant theme, string theory, dominated in the late 1970s and 1980s. Regge poles and S-matrix research had not fulfilled the promise that many of us felt they had in solving fundamental problems concerned with the nuclear forces.

Fortunately, my interests in quantum field theory while a student at Cambridge, and my publishing papers on the subject in the 1960s when I was at RIAS, made it possible for me to disentangle myself from the specialized subject of the analytic S-matrix and Regge poles, and to move into quantum field theory and new developments related to the weak interactions of particles. Many of my colleagues who had become specialists in S-matrix theory were not able to make the transition to the new fad, and the bells could be heard tolling for the end of their research careers.

This was another lesson to me about the brutality of physics: Modern physics develops through fads, often with many hundreds of physicists involved in the process. We can see this clearly in string theory today, where hundreds of string theorists at any given time are writing papers on string theory, and all citing one another's papers, which makes it appear that strings is the most important place to be in physics today.

This is in stark contrast to the past, when, for example, the revolution of quantum mechanics in the 1920s was accomplished by only a handful of theoretical physicists such as Wolfgang Pauli, Werner Heisenberg, Niels Bohr, Albert Einstein, Max Born, Erwin Schrödinger and Pascual Jordan. Also in contrast to string theory (a mathematical construct so far without experimental verification) and other speculative branches of research today, the creators of quantum mechanics paid close attention to experiments while developing the theory.

Although my physics was proceeding well, my marriage with Bridget did not survive the 1972 sabbatical. Later that year, I met an American physics graduate student, Margaret Buckby, at a conference. After Bridget and I separated, Margaret and I got together, and she moved to Toronto and entered the graduate school at U of T to complete her Ph.D. in experimental nuclear physics.

*

In 1974, I invited Murray Gell-Mann to Toronto for a week. Gell-Mann had won the Nobel Prize in 1969 for his many contributions to particle physics. He became well known for his 1964 paper introducing the quark model, "A Schematic Model of Baryons and Mesons," which revolutionized our understanding of the smallest constituents of matter. I had met Gell-Mann previously at conferences, but had not had the chance to spend much time with him. I went out to the Toronto airport to meet him. He entered the arrivals lounge toting a heavy suitcase, inordinately large for his physique. I offered to help him carry it, but he declined.

"What do you have in that bag, Murray?" I asked.

"This is my brain," he said. "I carry all my physics papers with me, and I have to prepare my talks while I'm in Toronto."

At the Four Seasons Hotel in downtown Toronto, I ushered him up to his suite, and then we proceeded down to the restaurant, choosing a corner table. Up until then, we had been discussing politics, the weather and Toronto traffic as I drove him to the hotel. We ordered something to eat, and suddenly the discussion switched to physics. Murray began to look tense, and the next thing I knew, he had planted his feet on my knees under the table! I was astonished; I felt as if I had suddenly slipped into a surreal parallel universe. Gell-Mann had unusually large feet for his stature, and his heavy brown leather shoes pressing on my knees was uncomfortable. What did this mean? And what was I to do? This famous physicist, father of the quark model, was my guest for a week at the university. There seemed to be two choices: I could tell him to take his feet off my knees, risking an embarrassing scene, or I could simply pretend that it wasn't happening, and continue talking about physics. Fortunately, I chose the latter option. After a while of talking physics, Murray seemed to relax in my presence. Possibly he felt that he had succeeded in dominating me intellectually. In any case, he removed his feet from my knees. Neither of us made any comment about the incident.

During the week, Murray displayed admirable professional behaviour. Although our particle physics group at U of T was small compared to the one at Caltech, where Murray was a professor, and other major centres for particle physics in the States, he put great effort into preparing a series of four lectures for the Physics Department students and professors. The lectures revealed a remarkable depth and breadth of knowledge of the subject, and were brilliant in their exposition. Gell-Mann reviewed all the important aspects of particle physics up until that time. His performance during these lectures clearly demonstrated why he was such a dominant figure in particle physics.

In the middle of the week, during one of his lectures, a secretary slipped into the seminar room and told me she had an important message for Professor Gell-Mann. Murray excused himself and left the lecture room for about fifteen minutes. When he returned, he announced that he had just had a phone call from a colleague at Caltech saying that the J/psi particle—in effect, the "charm" quark— had just been discovered independently at both the Stanford and Brookhaven accelerators. He then gave a spontaneous and brilliant discourse on this fourth quark. The J/psi resonance actually consisted of a charm quark and an anti-charm quark bound together. The name "charm" had been coined for the fourth quark in a paper by James Bjorken and Sheldon Glashow in late 1964. Murray also explained why the discovery of the charm quark was significant, and the role it played in the quark model that he had invented in 1964.

I already had some familiarity with this new quark. During a visit to the University of Wisconsin in Madison in 1964, where I attended a summer school in particle physics, I had written a paper proposing a fourth quark, constructed a fractionally charged quark model and extended it to include quantum spin and Gell-Mann's original three fractionally charged quarks. I didn't think up a sexy name for the quark—to follow the first three "flavours" called "up," "down" and "strange"—but I called it simply the fourth quark. The paper, opaquely titled "Higher Symmetries and the Neutron-Proton Magnetic-Moment Ratio," was published in 1965 in *Physical Review*. (Glashow and Bjorken's 1964 paper on the fourth, "charm" quark did not assign fractional charges to any of the quarks, and neither did other papers suggesting a fourth quark.)

One day during Gell-Mann's visit, we walked to a Hungarian restaurant on Bloor Street in Toronto. Over lunch, I asked him whether it was easier for him to publish papers now that he had a Nobel Prize and was world-famous. Murray became visibly upset and said, "Absolutely not! It's never easy! Every paper has its own problems."

He then told me the story about how he submitted his famous paper on the quark model to the *Physical Review Letters*, and it was rejected. He said that he was very angry about this, and even though it was late at night in Geneva, he phoned Leon van Hove, the director of the theory group at CERN, who was also an editor of *Physics Letters B* in Europe. Van Hove was not pleased about being phoned so late, and irritably asked Murray what he wanted. Murray said that he wished to submit a paper to *Physics Letters B*. Leon asked what it was about. Murray explained that it was about his idea that protons and neutrons were each made up of three particles. "What are these particles called?" Leon asked.

"They're called quarks," Murray answered, "named after the statement 'Three quarks for Muster Marks!' in James Joyce's *Finnegan's Wake.*"

According to Murray, there was a silence at the other end of the phone line. Then Leon asked, "What properties do these particles have?"

Murray answered that they had fractional electric charge. Again there was silence. Fractional electrical charges for particles were unheard of at that time—particles had charges of either +1 like the proton, -1 like the electron or zero like the neutron—and van Hove must have thought the idea was absurd. Murray asked van Hove whether it would be a good idea to submit a letter explaining this model to *Physics Letters B*. Van Hove did not think it was a good idea, Murray told me. However, Murray *did* submit the letter, it *was* published, and it became an important part of the reason that he was awarded the Nobel Prize in 1969.

As it turned out later, a young post-doctoral fellow at CERN, George Zweig, had independently discovered the same idea around the same time. He named the three particles making up the proton and neutron "aces," after the playing cards, and like Gell-Mann, he assigned them fractional charges. He also worked out important consequences of the quark/aces model for particle physics. Van Hove played an unfortunate role in this story, for when Zweig presented his paper to the CERN theory section for review, to get their permission to publish it in a journal, the permission was denied, and the paper remained unpublished for several years. One of the rules of the Stockholm Nobel Committee is that the Nobel Prize is only awarded on the basis of research papers published in peer-reviewed journals. However, the Nobel Prize was awarded to Gell-Mann not only for his celebrated paper on quarks, but for his contributions to the classification and interactions of elementary particles. Unfortunately, Zweig never received the Nobel Prize for his revolutionary contribution to the standard model of particle physics.

At a small party held to honour his presence at the University of Toronto, Gell-Mann and I were standing with a group of physics colleagues, and he suddenly said to me, "John, I understand you were born in Denmark."

When I confirmed this, he said, "Speak some Danish."

I made some benign comments in Danish, never thinking that Murray would understand what I was saying. But to my astonishment, he repeated what I'd said in Danish, and then corrected a word that I had used, saying that it didn't sound right to him.

I laughed and exclaimed, "Murray, I'm bilingual in English and Danish. How can you correct me?"

Then he asked me, "Where were you born in Denmark?"

"In Valby, on the outskirts of Copenhagen."

Gell-Mann exclaimed, "Aha! That explains it! I learned to speak Danish over a period of six weeks, living in Fredricksberg."

Fredricksberg is a district next to Valby, and indeed Fredricksberg Danish differs in subtle ways from the Danish spoken in Valby. I then realized that Gell-Mann was not only a genius in physics but also in linguistics. I had heard that he spoke several languages well and had the ability to pick up a language with unbelievable speed. In fact, after spending a week with him, I felt that Murray Gell-Mann was one of the most remarkable people I had ever met.

*

There were other famous visitors to the University of Toronto. In 1978, the committee organizing the annual Welsh Lectures, honouring our former chair, Harry Welsh, called for invitations for outstanding physicists to come and give public lectures. I proposed Sheldon Glashow and Abdus Salam. This was during the time of serious jockeying for the Nobel Prize honouring the development of the electroweak model, the theory that combined into one the electromagnetic force and the weak force. Both Glashow and Salam were becoming well-known for their work on this theory. They both accepted our invitation, and came as my guests together to Toronto for one week. They both gave review talks about particle physics and the recent developments in electroweak theory.

I had seen Salam in 1976, when early in that year I attended a high-energy physics conference at the University of Chicago. In a session on new developments in the electroweak model, I sat near the front of the auditorium next to my old supervisor, who wore his usual heavy overcoat and floppy fedora, even in the warm room.

I briefly told him about my new ideas on modifying Einstein gravity, based on Einstein's nonsymmetric field structure. Recently I had resumed working on gravity theory and relativity, along with my particle physics research.

Salam thought for a few moments and then said, "Well, John, Einstein's general relativity is so successful experimentally, it's hard to see how you could succeed with such a theory. But good luck with it anyway."

At that point we had to turn our attention to the speaker, Benjamin Lee, who was discussing recent results in developing an electroweak model of weak interactions. During the talk, he kept referring to the "Glashow-Weinberg model," and I could feel the tension rising in Salam. Every time this happened, Salam would clear his throat with a loud "Harrumph!" of indignation at the speaker not mentioning his name as an author of the electroweak model.

A few months later, I stopped at Imperial College on my way through London to visit the Rutherford Laboratory near Oxford, where I was conducting summer research with collaborators on scattering theory. I was in an office preparing my talk on pion-pion scattering, when the door burst open and Salam stormed in. He thrust a letter at me with a trembling hand and exclaimed, "John, read this!" The letter was very short: "Dear Benjamin, If you continue to ignore my work publicly and do not give me credit for developing the electroweak model, I will personally ruin your career." It was signed "Abdus Salam."

Astonished, I said, "Abdus, you cannot send this letter. It is simply unacceptable."

Salam tore the letter out of my hand, and said, "I'll send it if I want to. I refuse to put up with this. I deserve credit for my work!" He stormed out of the office, slamming the door behind him. I never discovered whether Salam sent that letter to Benjamin Lee. Tragically, Lee, his wife and child died in a car accident driving home from a physics laboratory in the United States not long after my conversation with Salam.

During his visit to Toronto in 1978, Salam seemed unusually concerned about my old 1964 paper on the fourth quark with fractional charges. The fourth quark played an important role in the development of electroweak theory. Interviewing me about it heatedly in my office one afternoon, he accused me of attaching the "wrong" electric charge of -1/3 to the fourth quark, instead of +2/3. He was upset with me about this. I had no idea why Salam was so exercised about this paper—and he was right about the incorrect electric charge—but later I thought that the paper might have played a minor role in the Nobel Committee's discussion about who was to get the prize, and who wasn't, for the electroweak theory. I still don't fully understand what the implications of this paper were.

Finally, in 1979, the Nobel Prizes began coming in for the electroweak theory. That year, the year after Glashow and Salam visited Toronto, the Nobel Committee awarded the prize for physics jointly to Glashow, Steven Weinberg and Salam for their development of the electroweak model and, in Glashow's case, for other contributions to particle physics. The prize was given for the electroweak theoretical model even though the particles that were crucial to prove the model, the W and Z bosons and the Higgs particle, had not yet been detected experimentally. The W and Z bosons were not detected until 1983, and Carlo Rubbia and Simon van der Meer, who led the team of experimentalists at CERN who found them, were awarded the Nobel Prize the very next year. At this writing, the Higgs particle has still not been detected.

*

My personal life took another turn in 1982, when I met Patricia Ohlendorf, who interviewed me for a science story on my modified gravity theory for a Canadian newsmagazine. With my own children, Sandra and Tina, then twenty-two and fifteen, I soon became a stepfather to Patricia's children, Derek and Tessa, who were then eight and six. Patricia and I were married in 1986.

In the years 1986 to 1993, I went through a burst of creative activity in my physics research. I ceased working on conventional calculations in particle physics, and developed riskier, more innovative ideas in cosmology, particle physics and gravity. I tended to adopt a contrarian position in physics, pushing against the current dominant paradigms and fads because I was curious about how robust they actually were.

Alan Guth, then of Stanford University, started one cosmology fad in 1981, called "inflation." This idea solved certain theoretical problems that attended the birth of the universe by having the universe suddenly inflate exponentially. There was no hard evidence to suggest that this had actually happened 14 billion years ago, and I wondered whether there was a theoretical alternative to the increasingly popular idea of inflation. I was pondering the so-called "horizon problem": various parts of the early universe were not causally connected because of the limit imposed by the constant speed of light, and this did not fit with the data from the cosmic microwave background (CMB), the evidence in the sky of the afterglow of the Big Bang, which appeared to be so uniform. This raised the following question: How could distant parts of the universe communicate with other parts, as they must have done for the CMB to be so uniform? Inflation did answer this question, by making the early universe prior to inflation much smaller than it was in the original Big Bang scenario, making it possible for light to traverse all parts of the universe.

It occurred to me in a flash that if the speed of light had been very much larger in the early universe, allowing distant parts to communicate with each other almost instantaneously, this would also immediately solve the horizon problem. This scenario meant, of course, that the speed of light would not be a constant, as everyone believed it was, and as postulated by Einstein in special relativity. Rather, the speed of light could vary over time.

Having a radical idea like this takes seconds, but to establish it as a theory requires a detailed mathematical development of the idea. I proceeded, over a period of several weeks, to develop this theory, which I called the Varying Speed of Light (VSL) cosmology, and I arrived at a new modification of Einstein's special relativity theory and gravitational theory, which was needed to support the idea of a very fast speed of light in the early universe. In the process, I went against the conventions of physics once again: in technical terms, I violated the symmetry of special relativity, the Lorentz transformation, which played a crucial role in the development of special relativity. I published my VSL model in 1993 in the *International Journal of Modern Physics D*, with the title "Superluminary Universe: A Possible Solution to the Initial Value Problem in Cosmology."

Another idea I worked on during those years, with my graduate student Darius Tatarski, was inventing a new model of cosmology that was based on an inhomogeneous solution of Einstein's gravitational field equations, first discovered by Georges Lemaître and Richard Tolman in the 1930s, and developed further by Hermann Bondi at Cambridge a decade later. Astronomers had observed large cosmic voids in space, areas containing very little matter and surrounded by filaments of galaxies. I found these voids to be profoundly significant in cosmology, and wondered if I could construct a cosmology model in which we were located within one of these voids. Tatarski and I found from our calculations that when light from distant galaxies passes through a void, it becomes dimmed and makes it appear that the galaxies are farther away than one would estimate them to be using the standard cosmology.* This occurs because the spacetime inside the void is expanding faster than outside, where most of the matter is present. We published this research in the *Physical Review* in 1992 and the *Astrophysical Journal* in 1995.

My void model actually predicted the startling observation in 1998 by two groups of astronomers, in Australia and the United States, that distant supernovae undergoing cataclysmic explosions appeared to be farther away than they should be according to the standard cosmology. That is, the light from the supernovae was surprisingly dim. The astronomers concluded that the expansion of the universe in the wake of the Big Bang was actually accelerating—that's why the supernovae appeared to be farther away than expected—and they suggested that the cause was what later came to be called "dark energy," which acted as a repulsive force driving the acceleration.

In my void model, the expansion of the universe is *not* accelerating, and there is no dark energy. The supernovae only appear to be receding, an illusion created by the void we inhabit. My model also predicted that Einstein's problematical cosmological constant is zero, and therefore has nothing to do with the apparent accelerating expansion of the universe. Most astronomers and physicists accepted the idea of dark energy quite readily in 1998, and the idea of a void cosmology was not popular. A major reason for this

---

*The standard cosmology is based on the Friedmann-Lemaître-Robertson-Walker (FLRW) spacetime geometry, which describes a homogeneous and isotropic universe.

was because the void model violated the so-called Cosmological Copernican Principle, which holds that our solar system, our planet, and we as human beings do not occupy a special, central place in the universe. In my void model, it is necessary that we as observers are near the centre of the huge void, which does put us in a special place in the universe.

However, what is known as the standard model of cosmology—based on a homogeneous solution of Einstein's equations, in which Einstein's cosmological constant produced the accelerated expansion of the universe—also suffers from a violation of the Cosmological Copernican Principle, but in terms of time, not space: the accelerating expansion of the universe supposedly commenced about 5 billion years ago, when our solar system was being formed. Given that the universe is about 14 billion years old, this seems a rather absurd coincidence.

Attempts to verify or falsify the Cosmological Copernican Principle have so far failed, because to falsify the principle, we would have to make measurements in other parts of the universe, which for obvious reasons is difficult. The violation of the Copernican Principle raises one of the most serious questions in modern cosmology: Do we, in fact, inhabit a special place in space or in time, or both? This obviously has significant implications for philosophy, religion and our understanding of our place in the universe.

In the late 1980s and early 1990s, exploring another topic altogether, I developed a new kind of relativistic quantum field theory, which was based on non-local fields. This was a highly complex and technically demanding endeavour. In standard quantum field theory, the fields such as the electromagnetic field and the fields associated with the particles of the standard model satisfy what is called "microscopic locality." That is, events associated with the particles and their fields that are relatively far away do not influence what happens locally. In contrast, my non-local fields at small scales *could* influence other physical processes not in the immediate vicinity.* This non-locality can be tested by experiments at the Large Hadron Collider at CERN. I published a paper on this new relativistic quantum field theory in *Physical Review* in 1990.

To satisfy my curiosity about how robust standard paradigms were in particle physics, I also constructed a model of electroweak interactions that did not include the elusive and undetected Higgs particle. Virtually all physicists believed that the Higgs field or particle was necessary to impart masses to all other elementary particles; they fully expected the Higgs particle to be detected eventually in high-energy accelerators. In my model, I only included particles that had been experimentally observed in high-energy accelerators at the time. I did, however, include the "top quark," which had not yet been detected in 1990, but was a few years later. My first paper on this research was published in a European journal, *Modern Physics Letters*, in 1991.

Because particle physicists had almost universally accepted the existence of a Higgs particle, my attempts to produce a model without one, even though the Higgs particle had not been detected, was a radical proposal that was not popular with other physicists. In my alternative model, the masses of elementary particles were generated at the quantum level, not by a classical scalar field like the Higgs field. I consider my alternative electroweak theory to be one way that the standard electroweak model could have been envisioned in the first place, back in the 1960s and 1970s. But the original theorists—Glashow, Salam, Weinberg and others—went off in

---

*An important consequence of my new field theory is that it rid relativistic quantum field theory of the ugly mathematical infinities that plagued it. It also made standard renormalization theory more comprehensible. However, my work also opened up the possibility of a finite quantum field theory in which the non-locality plays a fundamental role. This would remove the need for renormalization theory in particle physics.

a different direction, and did not incorporate the ideas of quantum field theory as I did.*

In addition to all these outside-the-box projects, I continued to ponder the problem of modified gravity and published papers on my nonsymmetric gravitation theory, NGT, on my own and with many talented graduate students over the years. I had been working on this theory all my adult life, since I was nineteen, and had corresponded with Einstein about his attempts to construct a similar alternative gravity theory. NGT was not a unified theory, aiming to combine gravity with electromagnetism and the forces in particle physics, but was instead a purely gravitational theory. Since 1978–79, I had concentrated on modifying Einstein's gravitational theory out of intellectual interest alone. Not until 1995 did it occur to me that my modified gravity theory could explain the anomalous behaviour of stars moving in approximately circular orbits in galaxies.

Astronomers had known since the 1930s that the outermost stars in galaxies were moving several times faster than was predicted by Newtonian and Einstein gravity. To explain these observations, without changing the theories, physicists had conjectured that there must be large "dark matter halos" attached to galaxies, which increased the strength of gravity and made the stars move faster. In contrast to this, my modified gravity theory, which contains no dark matter at all, explains the faster movement of stars because the theory has an intrinsically stronger gravity at certain distance scales. I published this work on NGT in collaboration with

---

*In collaboration with Viktor Toth, I took up this work on an alternative electroweak theory without a Higgs particle again in 2008, basing it on my non-local quantum field theory. However, I later decided that it would be possible to formulate an electroweak model without a Higgs particle using only physical local fields that do not violate causality. This could be accomplished by developing a novel, finite formulation of quantum field theory.

my post-doctoral fellow Igor Sokolov in *Physics Letters B* in 1995.

I continued to develop the theory over many years, and by 2006 it reached its final version in a paper published in the *Journal of Cosmology and Astroparticle Physics* called "Scalar-Tensor-Vector Gravity" (STVG). I gave this theory the popular name "modified gravity," or "MOG." By now I have published many papers on MOG with my graduate student Joel Brownstein and my collaborator Viktor Toth. The latest, simplest version of MOG is able to explain a large amount of data from the solar system to the outer edges of the universe without postulating the existence of exotic dark matter. The theory opens up the possibility of not having a singularity, or Big Bang, at the beginning of the universe, and it may also change the concept of black holes.

These five major research topics all challenged existing fads and paradigms. Yet the dominant fads, from inflation to dark matter and dark energy, to the standard electroweak model with a Higgs particle, were all based on a consensus in the physics community, not on hard experimental evidence. My alternatives to all these models were not greeted with enthusiasm by most other physicists, even though no obvious mistakes could be found in my calculations, and even though my alternative theories fitted the existing data just as well as the dominant paradigms, and sometimes even better.

Sailing against the current of modern consensus physics, I have often had difficulties getting my papers on these outside-the-box topics published in peer-reviewed, establishment journals. This was not the case with my earlier, thoroughly mainstream research on S-matrix, Regge poles, scattering theory and other topics in particle physics. Most often the referees who read and judged my new papers were themselves members of the mainstream research establishment, and had invested much time and funds in the standard models such as string theory, dark matter, the Higgs model and the homogeneous dark energy standard model of cosmology.

I learned a lot about the sociology of science during those creative but difficult years, as it became obvious to me that modern physics develops by consensus and less and less by observational data. Often, when reading referees' rejections of my papers, the message that came through loud and clear was: "I cannot find anything wrong with this paper, but I do not like it, and I don't think that most physicists would be interested in these ideas."

When I submitted my first paper on VSL, my alternative to inflation, to *Physical Review* in 1992, the referees chosen by the journal were scathing. They disliked the fact that I had modified the role of special relativity in Einstein's gravity theory to accommodate such a large speed of light in the very early universe, and they scorned my heretical suggestion that the speed of light was not a constant. They rejected the paper. I submitted it to a lesser-known European journal, the *International Journal of Modern Physics D*, which accepted it without any revisions, and it was promptly ignored.

Six years later, two other physicists, João Magueijo and Andreas Albrecht at Imperial College London, came up independently with the same idea of a varying speed of light. They submitted their paper to *Physical Review*, fought with the referees and editors for months, until finally the paper was accepted. Since then, my original paper has come out of hiding, and the idea of VSL, though it has not yet replaced inflation as the dominant early-universe scenario, is at least an active area of research.* Sometimes, one can simply be too early with a promising, non-mainstream idea.

---

*In 1998 and later in 2000 and 2001, I published in collaboration with my student Michael Clayton an elegant reformulation of the VSL theory called bimetric gravity theory. In this theory, the massless photons and gravitons have separate geometrical descriptions in spacetime, connected by a field. This allows for a larger speed of light in the early universe without violating Einstein's special relativity. Recently, Magueijo and collaborators have investigated this model and have found that it can agree with current observational cosmological data and produce predictions that can distinguish it from inflationary models.

The fate of my MOG paper in 1979 has an even more unusual and positive twist. When I first submitted "A New Theory of Gravitation" to *Physical Review*, it was rejected four times by four different referees. They claimed that there was no experimental need for a modification of Einstein's celebrated gravity theory—which is also what Abdus Salam had intimated to me that morning in Chicago. Also, the referees emphasized that Einstein's theory was elegant, and any modification would mar the beauty of the theory.

I decided to submit the paper for a fifth time. However, this time I did not receive an outright rejection, but the editor mailed me a manuscript that was a rewritten version of the manuscript I had submitted. The anonymous referee had checked every equation carefully, and had made some changes to the mathematics. At first I was upset to think that the referee had had the temerity to rewrite my whole manuscript without my permission. However, I calmed down after some days, and realized, prudently, that I now was guaranteed publication of the article in *Physical Review* if I agreed to accept the new version.

I wrote to the editor requesting the name of the anonymous referee, but he refused to divulge his or her identity. When the paper was finally published, I had to admit that the ghost-written version was an improvement over my original one. Over the years, this article has become one of my most cited ones, as today, thirty-five years later, modifying Einstein's gravity theory has become an increasingly popular research topic. I still feel a debt of gratitude to that anonymous referee, who had the courage to go against the grain of consensus physics, and I still wonder who this person was.

It wasn't only the peer-reviewed journals I often had difficulties with, but sometimes I was subjected to the scorn of colleagues and even my friends. For example, when my student Darius Tatarski defended his Ph.D. thesis at the University of Toronto on the cosmological void model, the examining committee, composed of my peers in physics and astronomy, ridiculed the whole idea. They claimed it was preposterous that we as observers should occupy such a special place in the universe, near the centre of a void. However, because the papers were published in respectable peer-reviewed journals, and in view of the quality of the research, the committee had no alternative but to award Tatarski his degree.

My new relativistic quantum field theory, too, provided good target practice. In early 1990, I invited Richard Woodard, a professor at the University of Florida at Gainesville, to give a series of lectures on string theory at the University of Toronto. Woodard had turned violently against the current fad, string theory, after studying the theory closely and publishing on it for several years, and I thought it would be interesting for our department to consider Richard's radical ideas.

Richard and I had lunch one day during his visit at the graduate school cafeteria. Over soup and sandwiches, he asked me what research I was doing at the time. I proceeded to explain my ideas on non-local quantum field theory. I had managed to get a paper on it published in the *Physical Review* in 1989, and was busy writing a follow-up one. Woodard is a volatile person, subject to outbursts when he feels provoked by physics he disagrees with or feels is simply wrong. "John, this is absolute nonsense!" Richard shouted at me. The graduate students around us at their tables began staring at these two older professors creating a ruckus, for I countered Richard by loudly defending my ideas. I considered Richard to be a brilliant theoretical physicist with awesome technical skills in performing complicated physics calculations. So his immediate negative response to my work—even though it was quite far outside the mainstream box of physics—caused me concern.

Yet a few weeks later, I received an e-mail from Woodard, announcing that he had had an "epiphany." He had seen the light, realized that my idea about non-local quantum field theory could lead to significant progress in particle physics and proposed that we collaborate. I was delighted, and readily agreed. In the meantime, my second, and seminal, paper on this subject, "Finite Nonlocal Gauge Field Theory," had been accepted by *Physical Review* and published in 1990. Woodard and I, along with my post-doctoral fellow Dan Evens and Woodard's graduate student Gary Kleppe, then produced a long and highly technical paper, which was published in *Physical Review* in 1991, applying my ideas on non-local quantum field theory to quantum electrodynamics.

It is clear from the history of science that real progress in understanding the secrets of nature is often only made through opposing the prevailing paradigms. From Copernicus and Galileo hundreds of years ago to more recent examples—Einstein's development of general relativity, Gell-Mann and Zweig's invention of the quark model, even the original ideas by Peter Higgs and others providing the basis of a standard electroweak model—we know that original ideas initially meet with skepticism. It is almost a given that new ways of seeing nature face severe opposition. This is not entirely a bad thing because a paradigm in science should not be changed until a new one has been thoroughly tested over time and survives. There's a built-in conservative attitude in scientific research, which is as it should be.

Still, it is encouraging to me to see some of my areas of non-mainstream research move a little closer to acceptance, and to see lively areas of research where there was initially scorn. For example, there are now many physicists working on alternative theories of gravity because dark matter and dark energy have become two of the most important problems in modern astrophysics and cosmology. To an increasing number of physicists, the idea that 96 percent of the matter and energy in the universe is invisible is becoming quite troublesome.

Another example is my violation of the symmetries of Einstein's special and general relativity, which was necessary in creating my VSL cosmology model. Today, violating these symmetries has become a major industry in physics. At a recent workshop at the Perimeter Institute, a physicist was lecturing on this very topic. I put up my hand and asked him, "Why are you violating the symmetries of special relativity?" He answered, "Isn't everybody doing it?"

*

In 1998, I was required to retire from the Physics Department at the University of Toronto, as the university had a mandatory-retirement policy then. I wasn't happy about being forced to retire, both for financial and psychological reasons. I could not imagine stopping my research. The word "retirement" was anathema to me, and still is. Physics for me is as essential as breathing, and life could not continue without my being fully involved in my research. However, I found that not having to teach courses and sit on committees freed up a lot of time for research. I still had my research grant from the Canadian government, and put it to good use travelling to conferences and supporting graduate students or post-docs.

Fortunately, in the same year that I retired, the Perimeter Institute for Theoretical Physics was getting started in Waterloo. It was aiming to become an institute that would conduct research on fundamental physics, much as the giant aerospace and defense company Martin Marietta had established my old fundamental-research institute RIAS in Baltimore decades earlier. I was invited by a colleague, George Leibbrandt, who was on the original board of directors of PI, to be either a member of the scientific advisory council or to participate in the research program. For me this was not a difficult decision. I was excited at the possibility of continuing my physics research in what looked like it could become a cutting-edge research environment.

For the first few years, I commuted to Waterloo from Toronto or from the island in the Kawartha Lakes region that Patricia and I had bought in 1995. But eventually it made more sense to move, which we did in 2003, buying a townhouse in north Waterloo across the road from Mennonite farm fields, a dramatic change from my nearly forty years in downtown, highly urbanized Toronto.

EPILOGUE

O NE MOVES ALONG IN LIFE, accepting as natural circumstances that to an observer might seem very strange indeed. As I matured in my career, I became focused on my teaching, research and administrative duties, without giving much thought anymore to the bizarre way that I had entered physics. Yet, during the writing of this memoir, I have had to revisit the past and have felt grateful all over again to the great physicists who helped and encouraged me when I was young.

Foremost was Einstein, who simply by writing to me and treating me as an intellectual equal helped me enter academic life and pursue a legitimate path towards becoming a professional physicist and teacher. Our exchange of letters and his views about physics at the end of his life greatly influenced my personal approach to research. Einstein was my first mentor in physics, and I feel that I have been continuing the work that he began and struggled with during his later years for more than fifty years now.

I also owe a significant debt of gratitude to Niels Bohr, Erwin Schrödinger and Paul Dirac for opening doors for me into academia and helping me obtain the financial support I needed for my research. The only "giant" of twentieth-century physics whom I met and write about in this memoir who was less than generous and helpful in my budding career was J. Robert Oppenheimer. Fred Hoyle was open-minded enough to take on a Ph.D. student with my anomalous background. Abdus Salam helped me succeed in earning my doctorate at Trinity College and was confident enough in my abilities that he hired me as his first post-doctoral fellow at Imperial College London.

But more important than these instances of practical help were the lessons I learned from these great physicists. They all, as well as Wolfgang Pauli, demonstrated how to be open to one's own new ideas in physics, how to do the hard work of formulating those ideas into mathematically rigorous models or theories and then how to defend those ideas and theories when the inevitable criticism from one's peers rains down.

Some of my mentors I never saw again after their initial positive influence on my life, such as Bohr and Schrödinger. Others, like Dirac and Pauli, I met again over the years. Of all the later meetings with these giants of twentieth-century physics, my last visit with Abdus Salam is particularly poignant. During a summer trip to Europe in the mid-1990s, I decided to visit my old professor at his International Center for Theoretical Physics in Trieste, Italy. One of Salam's important contributions to physics, aside from his research on the electroweak theory that earned him the Nobel Prize, was establishing this centre for theoretical physics as an intellectual home for scholars from developing countries in Africa, the Middle East and the Indian subcontinent. It had taken Salam's passion for physics and scholarship, and his concern for young researchers who were not able to advance as easily as those in Europe and North America, to establish this institute and obtain the funding necessary to build it and keep it going. I knew that Salam had been diagnosed with a rare form of Parkinson's disease, and I wanted to see him one more time while I had the chance.

The institute that had been built in 1968 in the Miramare Park had not changed over the ten or so years since I had last visited. It was an attractive complex, with offices for physicists, a library and space for the administrative staff. Salam had a private dwelling on the grounds, where his second wife, an English biologist, visited him frequently; his first wife travelled between Pakistan and London. The park and institute were located near the main road leading into Trieste, which itself was a charming relic from the Austro-Hungarian Empire.

Salam had a large office on the main floor of the institute. A plaque on its door announced that here was Professor Abdus Salam, Director of the Center. At the time of our appointment, I knocked on the door and a faint voice responded, "Come in." Upon entering the office, I was surprised to see large, dark footprints painted on the floor, leading from where Salam sat at his desk to the door I had just opened. As we greeted each other, I saw that he had grown much older. His once-gleaming black hair and moustache had turned grey and he looked frail and gaunt. Salam ushered me in with a weak wave of his hand, and I sat down in a chair near him. His arms and hands shook visibly, and he had difficulty speaking, but he managed to rouse himself for my visit. I now understood the purpose of those odd footprints on the floor. They were to help guide him through his office, from the desk to the door. They seemed a dark portent to me.

Salam looked at me gravely and asked, "John, how is your health?" I found this a surprising opening after so many years of not seeing each other, but of course his own health must now be the overriding factor in his life.

"I'm all right," I said.

"You are fortunate you are in good health," he said heavily. "As you can see, I am not doing so well."

We talked about the centre for a while. I wanted to broach the subject of its future, and in particular, who might be taking over when he was no longer able to function as director. However, I felt it would be insensitive to raise this issue, and instead we talked about the daily activities at the institute and what research was being done.

Salam then started discussing his own research, and we got round to talking about his Nobel Prize and his research on the electroweak theory. He said in a somewhat angry tone, "You know, John, after all that's been done, people don't believe me. They don't believe my theory."

He seemed upset, and it appeared that he was even upset with me. I realized that he may have read a recent paper I had published, in which I had redesigned the electroweak theory without using a Higgs particle, which was central to his and Weinberg's model.

"Do you believe in my electroweak model?" he asked.

"Well, Abdus," I began, "I think the basic structure of the theory is correct, particularly now that we know that the W and the Z bosons exist, and there is a need for the neutral Z particle, since they discovered neutral currents at CERN. However, I'm skeptical about the existence of the Higgs particle. As you know, it has some serious problems attached to it, such as the mass hierarchy problem and the cosmological constant problem."

He looked downcast and asked accusingly, "So you really don't believe in my model?"

"Well, I don't know," I said hesitantly. "In the end, experiment will decide whether or not the Higgs particle exists. However, I'm afraid that we're more than twenty or thirty years away from solving the mystery of what breaks electroweak symmetry and gives the W and Z bosons and fermions their masses. It might be the Higgs particle, but it might not be."

Even as I said this, I realized sadly that Abdus was never going to know the answer to this question, which had been so central to his life and work. I wondered whether I myself would live long enough to know the answer. Due to the exorbitantly high cost of building high-energy accelerators able to answer these questions, it would be a long time before these mysteries could be resolved.

"Unfortunately," I continued, "these days are not like the old days back in the fifties and sixties, Abdus, when we could produce a speculative theory, make a prediction and have it tested experimentally within a few months."

I then related to him an anecdote about Dick Feynman that I remembered from the sixties, when we were still trying to understand the structure of weak interactions of elementary particles. The U.S. physicists Robert Marshak and George Sudarshan discovered that the weak interactions had to have a vector minus axial-vector (V-A) structure. This technicality has to do with the kind of interactions that occur between elementary particles undergoing radioactive decay. Feynman and Gell-Mann independently discovered this V-A structure at Caltech at around the same time as Marshak and Sudarshan. Feynman invited his experimental colleagues to his office one day and excitedly told them that the weak interaction theory had to have this V-A structure: it was elegant and beautiful, so it had to be right! The experimentalists demurred, and told Dick that they had already checked all this out experimentally, and that the V-A proposed by Feynman and Gell-Mann was not the experimentally correct structure.

Feynman said, "Go away and do your experiments again. It's got to be correct."

Some months later, his experimental colleagues returned to Feynman's office and exclaimed, "Dick, you were right! We redid the experiments, and it is V minus A!" It did not take long to vindicate this prediction experimentally, compared to the situation in particle physics today, when it can take decades of effort to verify or falsify a theoretical prediction.

Salam continued to look downcast. I was sorry that I wasn't cheering him up during this visit. "You're right," he said. "The world of particle physics has changed profoundly. It may be that there are some fundamental questions that we can never find the answers to."

At the end of our conversation, I rose, held his trembling hand and wished him farewell and all the best. He nodded but said nothing. I followed the footprints on the floor, and at the door turned and waved to him with a heavy heart.

Salam died two years later. It turned out that, like so many successful founders of institutes and businesses, he had never groomed anyone to become his successor as director. Yet, after an initial confusing period of time, a new director was found, and the centre continues to fulfill its original purpose today, as the Abdus Salam International Center of Theoretical Physics. As for Salam's legacy as a physicist, which bothered him during our conversation near the end of his life, we now await confirmation of his electroweak theory by the Large Hadron Collider, or evidence that the Higgs particle does not exist and therefore the electroweak theory has to be revised.

*

I am now in my seventies, several years older than Salam was at our last meeting, and my joy in doing physics has not diminished. Thinking about how the universe works and how it can be explained by mathematical equations still astonishes and inspires me. My powers of reasoning and the flow of ideas seem to have only strengthened over the years, as so much physics has poured through my brain. I characteristically think laterally, pulling in information and insights from one area of physics or astrophysics to solve a problem in another area. This may help explain why I so often work outside the mainstream box in physics, exploring new ideas.

Looking back on the more than 250 papers I have published in peer-reviewed journals during my career, many were applications of fads and mainstream ideas that have long been forgotten. Who works on Regge poles today? The most important research I have published, which has received the most citations and the most attention within the physics community as well as in the media, has dealt with original ideas often well outside the mainstream of physics, and usually not following the latest physics fad at all— work such as my modified gravity theory (MOG), which has no dark matter, the varying speed of light cosmology (VSL), my void model of cosmology without dark energy and my alternative electroweak theory without a Higgs particle.

Often these papers were initially rejected by the established, peer-reviewed journals. Eventually I learned to gauge whether my ideas were possibly paradigm-shifting by noting the intensity of the negative opinions expressed by the referees and the loudness of the opposition of the physics establishment to my papers. There is a herd instinct in physics, as in most academic and scientific fields. Many hundreds of physicists are waiting for some exciting development within their mainstream activity, and so, when a prominent mainstream physicist publishes a new paper, the herd surges forward in that same direction, even though the idea underlying the new physics fad may not yet be properly thought out or has little possibility of experimental confirmation.

My collaborator Viktor Toth commented recently that there seems to have been an inversion in the age group taking risks and pursuing ideas well outside the mainstream of physics versus staying within the confines of the most recent fad. It used to be that the younger physicists in their day—those that I was privileged to know when they were middle-aged or older—such as Einstein, Heisenberg, Schrödinger, Pauli and Dirac—were the risk-takers. Quantum mechanics, one of the great revolutions of modern physics, was mainly developed by risk-taking young physicists. The older physicists involved in the development of quantum mechanics, such as Einstein, Bohr and Schrödinger, while they did contribute to the early creation of the theory, also played major roles as mentors. There was no draconian peer-review system during the early part of the twentieth century, when the radical quantum mechanics revolution was taking place. The older classical physicists, that is, the rest of the physics community, considered the new quantum mechanics to be incomprehensible and possibly crazy. However, the establishment did not impede the publication and dissemination of the major papers that were the groundwork of the new quantum mechanics.

Thus, unlike many younger physicists today, the younger physicists of the past had more freedom to take risks and publish cutting-edge research; they didn't worry so much that their papers would be rejected or that they would be denied tenure on the basis of their outside-the-box research interests. And, of course, quantum mechanics was developed by only about ten prominent physicists between 1920 and 1930, and the entire community of physicists numbered only several hundred. Today there are thousands of physicists actively pursuing physics research. Quantity does not necessarily equal quality.

Speaking to a young post-doctoral fellow at the Perimeter Institute recently and explaining my recent radical work in particle physics, he exclaimed that he was not in a position to attempt this kind of research because he had to worry about getting a job and achieving some sense of security so that he could support his wife and children. Another post-doctoral fellow I talked to said that it would be professional suicide for him to work on the type of ideas that I was investigating in particle physics. It should be understood that progress in physics does demand a continuing effort in mainstream physics. However, opportunities for creative young physicists are limited in that often physics departments and laboratories work in large groups, and you are expected to fit in with the group and pursue their goals, which are most often mainstream.

Today I am still as involved in discovering new ideas in physics and cosmology as I ever was. Perhaps the continuing surges of creativity that I experience might stem in some mysterious way from the psychic residues of my traumatic wartime childhood, or from my early ambition to become an artist, discovering and creating beauty. But I feel certain that this ongoing creativity is due in large part to my unusual path in physics. I was never subjected to the severe rigours of rote learning, which students undergo at universities as part of their early training as physicists, and therefore my creative abilities as a human being and a physicist were never quenched. Because I essentially taught myself physics and mathematics, I did not have to prove to the authorities at every step along the way that I was competent in successive areas of physics. My Ph.D. defence was the one exception, and it left a small psychic scar, as well as a trick stomach. But in general, when I was a student at Cambridge, as it was then in the 1950s, I was free to apply the tools I was learning to unlock the secrets of nature that compelled me.

I am grateful to Cambridge University for allowing me to go my own way. I am still enjoying the benefits, and battling with the difficulties, of journeying outside the herd. I am grateful to the generous giants of twentieth-century physics who served as my mentors and showed me the way.

Acknowledgements

I am much indebted to my wife, Patricia Moffat, for her superb editing of this book. Without her tireless efforts, it would not have been completed. I also thank Martin Green, João Magueijo, Pierre Savaria and Viktor Toth for their valuable comments and suggestions. I thank Martin Green particularly for suggesting the title, *Einstein Wrote Back*. Barbara Wolff at the Albert Einstein Archives at the Hebrew University of Jerusalem was very helpful in considering the treatment of Einstein's letters in this book and on the cover, and giving me permission to use quotes from and reproductions of the letters, for the university holds the copyright on all materials written by Albert Einstein, including personal letters.

I am grateful to my former agent, Jodie Rhodes, for her enthusiasm for this project from the beginning. Many thanks to Janice Zawerbny and Patrick Crean at Thomas Allen Publishers for their encouragement and helpful contributions throughout the writing of this book.

Index

Abdus Salam International Center of Theoretical Physics, 232
accelerating universe, 52n, 215, 216
accelerators, 52, 54, 75, 119, 123, 126, 164, 173, 182, 207, 217, 231
"aces" model, 209
Albert Einstein: Creator and Rebel (Hoffmann/Dukas), 168
Albrecht, Andreas, 220
Aldermaston Laboratory, 133, 134, 135, 145, 147
Amati, Daniele, 183
American Physical Society, 125, 165
Anderson, Carl, 120
angular momentum, 96, 96n (*See also* spin)
Annalen der Physik, 99
Annals of Mathematics, 87, 167
anthropic principle, 86
anti-matter, existence of, 120
anti-protons, 182
Astrophysical Journal, 215
astrophysics, 2–3, 52, 139, 199, 200
asymmetric field theory, 54
atomic bomb, 134, 145, 191, 198
atomic physics, 33–34, 39, 56, 75, 106, 183, 184, 184n
Avogadro's number, 100n

Bardeen, John, 137
Bargmann, Valentine, 193–94

Barrow, John, 70
Battle of Britain, 9–10
Bender, Welcome, 158–59, 191–92
Bergman, Otto, 163, 170, 172
Bergman, Peter, 88
Berkeley group, 183, 183n
Bern conference. *See* Einstein, anniversary conference
Bertotti, Bruno, 64
Bethe, Hans, 121–22
Bianchi identities, 126
Bianchi, Luigi, 126
Big Bang model, 48, 78, 79–80, 84, 85, 90, 212, 214, 215, 219
Bimetric gravity theory, 220n
Bjorken, James, 207, 208
black holes, 48, 90, 171–72, 171, 219
Blackett, Patrick, 198
Bogoliubov, Nikolay, 142
Bohr, Niels, 4, 31, 33–42, 59, 60–61, 66, 67–68, 70, 113, 120, 127, 130, 132, 180, 205, 227, 228, 234 (*See also* Einstein–Bohr discourses; Niels Bohr Institute)
bombs, 9, 10, 11, 12, 15, 18, 127, 129, 145, 127, 145, 191, 192, 198
Bondi, Hermann, 78, 79–80, 93, 102, 214
Bonnor, William, 61, 63, 69, 95–96, 103, 146, 147, 148
Born, Max, 101, 110, 205

Bose, Satyendra Nath, 96, 102
Bose-Einstein statistics, 97
bosons, 96–97, 122n, 135–36
"bra" and "ket" notation, 124
Brandeis University, 95
Brans, Carl, 70
Bransden, Brian, 183–88, 189, 191, 195, 199
Bristol bombings, 8–9, 12, 120
British Atomic Energy Laboratory, 133
Brownian motion, 100n
Brownstein, Joel, 219
Buckby, Margaret, 205
Burbidge, Geoffrey, 74–75, 84
Burbidge, Margaret, 74–75, 84
Buridan, Jean, 51, 51n

Caltech, 74, 98, 120, 207
Cambridge University, 2, 4, 62–91, 93–94,
    105–17, 109–32, 137–38, 139–49, 198, 202,
    204, 214, 235
"Can Quantum-Mechanical . . ."
    (Einstein/Podolosky/Rosen), 100
Canadian Association of Physics, 201
Canadian Journal of Mathematics, 148
Candlin, David, 114, 144
carbon resonance, 85–86
Carter, Brandon, 70
Cavendish Laboratory (Cambridge), 119, 128,
    137, 141
Cavendish Maxwell Lecture Theatre
    (Cambridge), 127, 128
CERN, 3, 119, 152, 177, 179–84, 192, 195, 208,
    209, 217
Chandrasekhar, Subramanyan, 86
Chantilly conference, 169–72
"charm" quark, 207–8
Chew, Geoffrey, 183, 187
civil rights movement, 162–63, 199
classical physics, 34, 50–51
cold war, 162, 180, 192, 199
Coleman, Sidney, 166
colour theory, 66
Columbia University, 116, 129–30, 155, 173
Communist Party, 198
complex variable theory, 158
computers, 3, 184, 185

condensed matter physics, 89
Copernicus, 216, 223
Coral Gables conferences, 135–36, 169–70
cosmic microwave background (CMB),
    213–14
"Cosmological Considerations in . . ."
    (Einstein), 53–54, 53n
cosmological constant, 53, 53n, 57, 215, 216,
    230
Cosmological Copernican Principle, 216
cosmology, 25, 27, 48–49, 52–53, 52n, 69–70,
    214–16
covariance principle, 47n
Cuban missile crisis, 199
Curie, Marie, 137
cyclotron, 182

dark energy, 52, 52n, 53, 57, 215–16, 219, 223
dark matter, 52, 52n, 57, 218, 219, 223, 233
Dept. of Scientific and Industrial Research
    (UK), 59, 60, 62, 95, 133, 134, 145, 146–47,
    159
Deser, Stanley, 95
determinism vs. non-determinism, 50–51
Dewasne, Jean, 23
Dewitt, Bryce, 134–35, 170
Deyrolle, Jean, 23
Dicke, Robert, 70
Dirac equation, 120, 123
Dirac, Manci, 133, 137–38
Dirac, Paul, 4, 62, 70, 81, 86, 111–13, 116–17,
    119–38, 142–43, 144, 145, 161, 169–70, 175n,
    227, 228, 233
dispersion relations, 123, 141–42, 158, 183, 196,
    197
DNA, 67, 122
"Do Gravitational Waves Exist?"
    (Einstein/Rosen), 98–99, 101
"Does the Inertia . . ."(Einstein), 99n
Dublin Institute for Advanced Studies, 62, 64,
    67
Dubna accelerator, 119, 180
Dukas, Helen, 168
dynamic evolving model, 80 (See also Big
    Bang model)
Dyson, Freeman, 142

$E=mc^2$, 99n, 182n
Eddington, Sir Arthur, 25, 27, 171
Eden, Richard, 144
Ehrenfest, Paul, 116
Einstein, Albert, 4–5, 28–31, 94, 233, 234
　anniversary conference (Bern), 93–99,
　　101–4, 148, 191
　awards, prizes, 55, 55n
　– Bohr discourses, 37–40, 48, 49, 56–57, 100
　detractors, 43, 64, 67–68, 76, 88, 113
　– Infeld paper, 148, 166–68
　– Kurşunoğlu correspondence, 169
　– Moffat correspondence, 4, 43–55, 60, 77,
　　94, 104, 218, 227
　(*See also specific theories, papers, books, etc.*)
Einstein, Hans Albert, 104
Einstein, Mileva, 104
Einstein-Podolsky-Rosen (EPR) Paradox,
　100, 164
Einstein-Rosen Bridge, 100
electroweak theoretical model, standard, 210,
　211, 212, 219, 223, 228, 230, 232
electroweak theory, alternative (Moffat),
　217–18, 218n, 230
electromagnetism, 5, 28, 30, 38, 39, 43, 54, 56,
　57, 62, 67, 68, 78, 97–98, 117, 131, 153, 176, 218
　(*See also* unified field theory)
electron volt, 182, 182n
electrons, 34, 66, 66n, 96n, 106, 116, 120, 123,
　125n, 130, 131–32, 153, 182, 209
elementary particles, 96n
elements, origin of, 84, 96n
Ellis, George, 70
ether concept, 29
Evens, Dan, 223
Evolution from Space (Hoyle), 86
Evolution of Physics, The (Infeld/Einstein), 56
exclusion principle, 96n, 106
expanding universe, 52n, 53n, 79, 214–16
experiment/theory interplay, 3, 52
experimental physics, defined, 3

Faraday, Michael, 29
Feldman, Gordon, 164, 187
Fermi, Enrico, 96n
fermions, 96n, 136, 230

Ferranti Mercury computer, 184
Feynman, Richard, 156, 231
field equations, 47n, 48, 53–54, 55, 63, 67, 68,
　87–88, 90, 98, 101, 124, 125n, 126, 131, 153,
　171, 214
field theory, 29–30, 48–49, 53 (*See also* quan-
　tum field theory)
"Finite Nonlocal Gauge Field Theory"
　(Moffat), 223
first law of mechanics (Newton), 86
Flexner, Abraham, 162n
Florida State University at Tallahassee, 135
Flowers, Bridget, 83, 91, 132 *(See also* Moffat,
　Bridget)
FLRW spacetime geometry, 215n
Fock space, 101–2
Fock, Vladimir, 101, 103
"Foundations of a Generalized . . ." (Moffat),
　79
fourth quark, 207–8, 212 (*See also* "charm"
　quark; J/psi resonance)
Fowler, William, 74–75, 84, 86
fractional electric charge, 207, 208, 209, 212
Fubini, Sergio, 183
Fulton, Robert, 164

galaxies, 25, 52n, 57, 79–80, 85, 214–15, 218–19
Galileo, 29, 223
Gamow, George, 53n, 80
Gatland, Ian, 153, 164
gauge theory, 97
Gell-Mann, Murray, 156, 166, 197, 204, 206–10,
　223, 231
general relativity theory, 25, 28, 29, 47–48,
　47n, 52n, 163, 176, 211, 223, 224 (*See also*
　gravity theory, Einstein)
Gestetner instrument, 186–87, 195
Gibbons, Gary, 70
Gilbert, Walter, 122, 140
Glashow, Sheldon, 207, 208, 210, 212, 217–18
Glenn L. Martin Company. *See* Martin
　Marietta
Gödel, Kurt, 162n
Gold, Thomas ("Tommy"), 78, 79–80
Goldberger, Marvin ("Murph"), 174–75, 176,
　192, 193

Goldstone, Geoffrey, 121–22, 122n
Goudsmit, Samuel, 75, 116
"Gravitational Collapse and ..." (Penrose), 90
Grant, Cary, 95, 120 (*See also* Leach, Archibald)
gravitational waves prediction, 98, 99, 101–2, 103
gravity theory (Einstein), 4–5, 38, 44, 48, 49n, 52, 53, 54–55, 75, 75n, 87–90, 171, 176, 211, 221 (*See also* general relativity theory; modified gravity [MOG] theory; unified field theory
gravity theory (Newton), 69, 70, 75n
Green, H.S., 110
Greenall, Mr., 59, 60–62, 63, 64, 69, 77, 95
Guth, Alan, 213

Haag, Rudolf, 43, 114
Haag's theorem, 114–15, 124, 125
Hamilton, James, 105–6, 111, 113–14, 116, 121, 122, 123, 128, 130
Hartung, Hans, 23
Harvard University, 122, 153
Hawking, Stephen, 70, 90
h-bar, 129, 316
Heisenberg, Werner, 34, 51, 66, 106, 112, 127–32, 183, 205
helium, 84
Higgs particle, 152, 153, 212, 213, 217, 218n, 219, 230, 232, 233
Higgs, Peter, 152, 153, 223
Hilbert space, 123–24, 124, 134
Hlavaty, Vaclav, 168–69
Hoffmann, Banesh, 87, 166–68
homogeneity principle, 79, 215n, 216
horizon problem, 213–14
Hoyle, Fred, 71, 74–75, 78, 79–80, 81, 84–85, 93, 96, 139, 144, 228
Hoyle's fallacy, 86
Hubble, Edwin, 53n, 79–80
hydrogen, 84
hydrogen fusion bomb, 198

Il Nuovo Cimento, 88, 188
Illums Bolighus gallery, 24

Imperial College London, 95, 144, 147, 149, 151–59, 164, 211, 220, 228
industrial physics, 3
inertia, origin of, 70, 86–87
Infeld, Leopold, 48–49, 56, 87, 89, 101–2, 101n, 103, 148, 166–68, 173
infinities, cancellation of, 112, 112n, 175
inflationary models, 213–14, 219, 220, 220n
Institute for Advanced Study (Princeton), 44, 45, 49, 56, 88, 98, 104, 141, 167, 174, 194, 195–96
International Center for Theoretical Physics (Trieste), 137, 228–29
International Journal of Modern Physics D, 214, 220
invariance principle, 47n (*See also* covariance principle)
inverse scattering amplitude method, 182–83, 192
invisible matter, 52, 52n, 57 (*See also* dark matter; dark energy)
isotropy principle, 79, 215n

J/psi resonance, 207 (*See also* "charm" quark)
Johns Hopkins University, 164, 187
Jordan, Pascual, 70, 205
Journal of Cosmology and Astroparticle Physics, 219
Journal of Rational Mechanics and Analysis, 168

Källén, Gunnar, 180–81, 193
Källén-Lehmann representation, 173–74, 180
Karsh, Yousuf, 88
Kaufman, Bruria, 97–98
Kennedy, John F., 199, 200
Kerr, Roy, 87–88, 89, 147, 167
Khrushchev, Nikita, 199
Kibble, Tom, 154
King's College London, 93, 96, 102, 158
Kleppe, Gary, 223
K-meson, 125–26, 131
Korean War, 133, 161
Kronig, Ralph, 116
Kruskal, Martin, 171n, 172
Kundt, Wolfgang, 170, 172
Kurşunoğlu, Behram, 136, 137, 169–70

Large Hadron Collider (LHC), 3, 152, 177, 189, 217, 232

Lattes, César, 173n

Lazaridis, Mike, 1

Leach, Archibald, 120 (*See also* Grant, Cary)

Le Verrier, 75n

Lee, Benjamin, 211, 212

Lee, T.D., 156

Lefschetz, Solomon, 161–62

left-right parity symmetry, 155–56

Lehmann, Harry, 173–74, 180, 181

Leibbrandt, George, 224

Lemaître, Georges, 80, 85, 214

Lichnerowitz, André, 170

lithium, 84

Lorentz, Hendrik, 29, 49, 87, 116, 153, 214

Lovelace, Claude, 185–86

Mach, Ernst, 69–70

Macquarie University, 164

Magueijo, João, 220, 220n

Mandelstam, Stanley, 183, 183n, 187

Manhattan Project, 191, 198

Marshak, Robert, 201, 231

Martin Marietta (co.), 162, 191–92, 224

Massey-Harris-Ferguson Foundation, 77

matrix mechanics, 66, 66n, 132

Matthews, Paul, 90, 139, 141, 143–44, 154, 158, 159

Max Planck Institute, 131, 132

Maxwell, James Clerk, 29, 30, 54, 56, 125n, 138, 153, 176

McCarthy era, 198

McCarthy, Ian, 106–11

McCrea, William, 61, 63, 69, 146, 147, 148

Meaning of Relativity, The (Einstein), 45

Meggs, John, 143–44

Mercier, André, 96, 102

Mercury, perihelion advance of, 75, 75n

microscopic locality, 216

military conscription, 133, 134, 161

military contracts, 162, 192

Modern Physics Letters, 217

modified gravity (MOG) theory, 57, 78, 79, 87, 94, 144, 147, 148, 213, 218–19, 220, 221, 233

Moffat, Bridget, 146, 147, 149, 165, 162, 168, 202 (*See also* Flowers, Bridget)

Moffat, Christina ("Tina"), 203, 213

Moffat, Esther, 7–16, 27, 136, 190

Moffat, George, 7–16, 19–20, 23–24, 26, 27, 30–31, 90–91, 146, 148, 189–90

Moffat, Patricia, 225 (*See also* Ohlendorf, Patricia)

Moffat, Sandra, 177, 188, 199, 213

molecules, size of, 100n

Møller, Christian, 42, 43

motion of particles, 44, 49, 54, 56, 80, 87–88, 103, 144, 148

Mott, Neville, 128, 129, 130

Nature of the Physical World, The (Eddington), 25, 27

Nazism, 67, 127, 129, 170, 174

neutrons, 173n, 183, 208, 209

"New Theory of Gravitation, A" (Moffat), 221

Newton, Isaac, 28, 69, 74, 86, 119, 138, 166

Niels Bohr Institute, 31, 33, 43, 44, 46, 57, 59, 114, 180

Nobel Prize, 34, 55, 62, 86, 106, 120, 121, 122, 128, 132, 133, 137, 153, 155, 156, 193, 206, 208, 209, 212–13, 228, 230

nonsymmetric unified field theory, 39, 55, 98, 99, 168, 169, 211

nonsymmetric gravitation theory (NGT), 218–19 (*See also* modified gravity, MOG)

nuclear forces, 54, 55, 173n

Nuclear Physics, 173

nuclear physics, 56, 75, 110, 121, 191, 193

nuclear weapons, 162, 192, 199

nucleons, 131, 173

Nuffield Foundation, 77, 90

Ohlendorf, Derek, 213

Ohlendorf, Patricia, 213 (*See also* Moffat, Patricia)

Ohlendorf, Tessa, 213

"On Spherically Symmetric Solutions ..." (Kerr), 88

"On the Electrodynamics of Moving Bodies" (Einstein), 97, 99

"On the Formulation of Quantized . . ."
  (Symanzik/Zimmermann), 174
"On the Integrability Conditions . . ."
  (Moffat), 168
"On the Motion of Particles in . . ."
  (Einstein/Infeld), 88
Oppenheimer, J. Robert, 56, 97, 102, 120, 142,
  191, 192, 194–99, 228
Oswald, Lee Harvey, 199
Otte, Henry, 164
"outside the box" physics, 1–2, 4, 213–24, 220n,
  228, 232–35
Oxford University, 164, 165

P-wave resonance, 184–93, 184n, 195 (See also
  rho-meson)
Page, John, 30, 31, 40, 42
Pais, Abraham, 196
Palmer Laboratory (Princeton), 191, 192
Papapetrou, Achilles, 87, 103
para-statistics field theory, 110
parity violations in weak interactions (See
  left-right parity symmetry)
particle physics, 50, 75, 90, 99n, 119, 121, 122,
  122n, 139, 140, 149, 147, 152, 158, 164, 177,
  179–89, 196, 202, 206–7
"Particle Problem in the General . . ."
  (Einstein/Rosen), 100
Pauli, Wolfgang, 34, 68, 96, 96n, 102–4, 105–17,
  127, 130, 132, 140, 155, 156–57, 176, 180, 193,
  205, 228, 233
peer review, 98, 99, 101, 187, 209, 219–20, 221,
  222, 233, 234
Penrose diagrams, 90
Penrose, Oliver, 89
Penrose, Roger, 89–90
perfect cosmological principle, 79–80
Perimeter Institute for Theoretical Physics
  (PI), 1, 224–25, 234
photino, 136
photons, 96–97, 96n, 125n, 220n
Physical Review, 75, 88–79, 98–99, 100, 169,
  182, 202, 208, 215, 220, 222
Physical Review Letters, 90, 187, 188, 189, 191,
  208, 217
Physics Letters B, 208, 209, 219

pi-mesons or pions, 131, 173, 173n, 182–89
pion-pion scattering, 173, 182–89, 183n, 184n,
  193, 211
Pirani, Felix, 74, 76, 80–81, 86, 89, 96
Pius XII (pope), 85
Planck, Max, 34, 129
Podolsky, Boris, 100, 164
Poincaré, Henri, 29
Poliakoff, Serge, 21, 22–24, 31
Polkinghorne, John, 121, 140
positrons, 120, 182
Powell, Cecil, 173n
Princeton University, 44, 45, 49, 56, 88, 90, 98,
  100, 104, 133, 141, 158, 161, 167, 171n, 174, 191,
  192, 195–98
Principia (Newton), 69
Principles of Quantum Mechanics (Dirac), 126
probability theory, 34, 47, 48, 51
Proceedings of the Cambridge Philosophical
  Society, 78–79, 82, 87, 94, 145
Proceedings of the Franklin Institute, 99
Proceedings of the Royal Irish Academy, 68
Proceedings of the Royal Society, 63
proton synchrotron, 182
protons, 34, 173n, 182, 183, 208, 209

quantum electrodynamics (QED), 125, 125n,
  135, 164n, 176, 180, 223
quantum field theory, 41, 47, 90, 96, 97, 101,
  114–15, 119, 123–24, 131, 134, 139, 141, 144, 152,
  163, 164, 173–76, 175n, 180, 183, 204, 205, 216
quantum field theory, non-local (Moffat),
  216-17, 217n, 218n, 222-23
quantum gravity, 1, 50, 52, 78, 112n, 135, 176
quantum mechanics, 34, 38, 47, 50–51, 81, 96,
  100, 106, 115, 120, 123, 128, 130, 132, 205,
  233–34
quantum physics, 99n
quark model, 173n, 206–9, 217, 223
quasars, 200
Queen's University (NY), 166

racism, 143–44, 162–63
radioactive decay, 155, 231
Rees, Martin, 70
Regent Palace Hotel (London), 60, 62

Regge poles, 164, 202, 203, 204–5, 219, 233
Regge, Tullio, 164
religion. *See* science and religion
Renner, Bruno, 204–5
renormalization theory, 112, 112n, 164, 164n, 175–76, 175n, 217n
Research in Motion (RIM), 1
resonance bump. *See* P-wave resonance
rho-meson, 189, 191, 195 (*See also* P-wave resonance)
RIAS, 158–59, 161, 170, 171, 173, 174, 182, 189, 191–92, 200, 205, 224
Riemann, Georg Bernhard, 78n
Riemannian geometry, 78, 78n, 94, 126, 147
Robertson, Howard Percy, 98, 101
Robinson, Ivor, 89
Rochester University, 201
Rosen, Nathan, 98–99, 100, 101
Rosenkrantz, Dr., 35–36, 37, 38, 40, 41, 42
Roughton, Alice, 81–83
Roughton, Francis, 81–83
Rubbia, Carlo, 212–13
Rutherford Laboratory, 211
Rutherford, Ernest, 33, 52, 141

Salam, Abdus, 90, 111, 137, 139–44, 147, 151–57, 164, 165, 166, 169, 210–12, 217–18, 228–32
Salon des Réalités Nouvelles, 23, 24
Sanderson, Mr., 77, 90–91
scattering theory, 114, 202, 204, 211, 219
Schild, Alfred, 80
Schladming lectures, 204–5
Schrödinger, Anny, 64–65, 69
Schrödinger, Erwin, 4, 39, 62, 63–69, 70, 76, 77, 113, 120, 124, 132, 205, 227, 228, 233, 234
Schrödinger wave equation, 66, 66n, 132
Schwarzschild solution, 90, 171
Schwarzschild coordinates, 172
Schwed, Philip, 163, 164
Schwinger, Julian, 41, 196
Sciama, Dennis, 69–71, 74, 76, 86–87, 89
science and religion, 84–85, 121, 140
Second World War, 7, 8–16, 18, 21, 22, 120, 127, 129, 170, 174–75
Serber, Robert, 197
sexism, 156, 165

Shirkov, Dmitry, 142
S-matrix theory, 183, 187, 195, 202–5, 219
Sokolov, Igor, 219
solar system, 3, 75, 216, 219
Solvay conference (1927), 48
Soulages, Pierre, 23
Soviet Union, 162, 180, 192
spacetime, 25, 28, 29, 49n, 67, 75n, 78, 100, 122, 136, 172, 173, 175, 176, 215, 215n, 220n
Spacetime and Gravitation (Eddington), 25, 27
Space, Time, Matter (Weyl), 97
special relativity theory, 28–29, 97, 99n, 122, 214, 220n, 224
spectral line radiation, 33–34, 33n
speed of light, 214, 220, 220n
spin, 96, 96n, 116, 120, 122, 135-136, 173, 182, 184, 208
St. John's College. *See* Cambridge University
Stanford University, 207, 213
stars, 25, 33n, 53n, 74–75, 78, 85–86, 90, 139, 171, 218–19
statistical mechanics, 66
steady-state model, 78, 79–80, 85
Steenberg, Dick, 202
Stephenson, Geoffrey, 95–96, 103
Streeter, Ray, 152
string theory, 50, 52, 176, 205, 219, 222
Structure of Spacetime, The (Schrödinger), 66, 67
subatomic physics, 97 (*See also* particle physics)
Sudarshan, George, 231
sun, 34, 75, 75n
"Superluminary Universe . . ." (Moffat), 214
supersymmetry, 135–36
Symanzik, Kurt, 174–76
symmetry breaking, 122, 122n, 224
Synge, Professor, 64
Szekeres, George, 171n, 172

Tatarski, Darius, 214, 221–22
Tate, John, 100
Taylor, John Clayton (J.C.), 152, 154
Taylor, John Geoffrey (J.G.), 196–98
Taylor, Pat, 196
Teller, Edward, 198

Templeton, Sir John, 121
theoretical physics, defined, 3
Theory of Quantized Fields, The
    (Bogoliubov/Shirkov), 142
Tolman, Richard, 214
top quark, 217
Toth, Viktor, 218n, 219, 233
Trieman, Samuel, 192, 193
Trinity College. *See* Cambridge University
tripos exams, 76, 79, 81, 141

Uhlenbeck, George, 116
uncertainty principle, 51, 129, 132
"Unified Field Theory" (Moffat), 54
unified field theory (Einstein), 28, 29–31, 30n,
    33, 36, 37–38, 42, 43–44, 49, 54, 61, 77–78,
    131, 169
unified field theory (Heisenberg), 112, 127–32
U.S. Air Force, 162, 169
University of Bern, 96
University of Bristol, 173n
University of Chicago, 211
University of Copenhagen, 27, 28, 41
University of Florida, 222
University of Graz, 67
University of Hamburg, 173
University of Indiana, 168–69
University of Liverpool, 61, 63
University of London, 146
University of Lund, 180–81
University of Miami, 135–36, 169
University of North Carolina, 170
University of Paris, 170
University of Texas (Austin), 134
University of Toronto, 89, 201–24
University of Warsaw, 102, 148
University of Wisconsin, 189, 207–8

V-A structure, 231
van der Meer, Simon, 212–13
variable gravitational constant G, 70, 218
Varying Speed of Light (VSL) cosmology, 214,
    220, 220n, 224, 233
Vasarely, Victor, 23
void cosmology, 214–16, 222, 233
von Hove, Leon, 187, 208–9
von Neumann, John, 162n
"Vulcan" (planet), 75n

W and Z bosons, 212–13, 230
Walker, W.D., 189
Ward identity, 164n
Ward, John, 164–66
wave mechanics, 66
Weinberg, Steven, 173–74, 212, 217–18, 230
Weisskopf, Victor, 106
Welsh, Harry, 201, 202, 203, 210
Welsh Lectures, 210–11
Weston-super-Mare, bombings, 10–11
Weyl, Hermann, 30, 65, 97, 102
Wheeler, John, 90, 142, 158, 171–72
Wightman, Arthur, 192–93
Wigner, Eugene, 133, 137, 192, 193–94
Witten, Edward, 163
Witten, Louis, 163
Woodard, Richard, 222–23
wormhole, 100
wranglers (tripos exams), 141
Wu, Chien Shiung, 155

Yang, C.N., 156
Yeshiva University, 125

Zimmermann, Wolfhart, 174
Zweig, George, 209, 223